Teaching Mathematics
A *Sourcebook of Aids, Activities, and Strategies*

SECOND EDITION

Max A. Sobel and *Evan M. Maletsky*

Montclair State College
Upper Montclair, New Jersey

Prentice Hall, Englewood Cliffs, New Jersey 07632

Library of Congress Cataloging-in-Publication Data

SOBEL, MAX A.
    Teaching mathematics.

    Includes index.
    1. Mathematics—Study and teaching (Secondary)
I. Maletsky, Evan M.   II. Title.
QA11.S663 1988   510'.7'12   87-10341
ISBN 0-13-894155-6
ISBN 0-13-894148-3 (pbk.)

Editorial/production supervision
  and interior design: Patrick Walsh
Cover design: Diane Saxe
Manufacturing buyer: Carol J. Bystrom

Printed in the United States of America

10 9 8 7 6 5 4 3 2 1

ISBN 0-13-894155-6
ISBN 0-13-894148-3 {PBK.}   01

Prentice-Hall International (UK) Limited, *London*
Prentice-Hall of Australia Pty. Limited, *Sydney*
Prentice-Hall Canada Inc., *Toronto*
Prentice-Hall Hispanoamericana, S.A., *Mexico*
Prentice-Hall of India Private Limited, *New Delhi*
Prentice-Hall of Japan, Inc., *Tokyo*
Prentice-Hall of Southeast Asia Pte. Ltd., *Singapore*
Editora Prentice-Hall do Brasil, Ltda., *Rio de Janeiro*

Photographs that appear throughout the book
were taken by Lorin P. Maletsky
and Kerry D. Maletsky

Contents

Preface

Teaching Mathematics: A Sourcebook of Aids, Activities, and Strategies is designed for use by the mathematics teacher and the teacher in training. It treats the art of teaching through a series of motivational ideas suitable for many grade levels and abilities, and through a discussion of activities, materials, and manipulatives suitable for classroom use in individual or class instruction.

The first edition of *Teaching Mathematics* was written in response to the requests of many teachers and students of mathematics who attended lectures, courses, and workshops given by the authors, but were unable to find appropriate materials in a single convenient source. The authors collaborated to provide a collection of teaching aids, activities, and strategies suitable for elementary and secondary mathematics classes. In this revised and expanded second edition, greater attention has been given to problem-solving experiences in the classroom and to organizing the material about specific subject matter areas.

Part I of this text consists of chapters 1 through 4. Here we address the general question of the art of teaching mathematics, with specific attention given to the role, importance, methods, and techniques of motivation. Many educators agree that motivation is the key to success for both teaching and learning mathematics, and that sufficient time must be taken in the daily activity of a class to be certain that students are suitably motivated. This point of view prevails throughout this second edition.

Problem solving emerged in the 1980s as a major concern for mathematics educators, and promises to capture our attention in the 1990s as well. A new Chapter 3 explores various strategies for problem solving, with special attention given to methods of motivating students in this important topic. This is followed by a chapter on mathematical recreations and enrichment topics that serve both to motivate and instruct, as well as provide for individual differences, offering a glimpse of some of the exciting aspects of mathematics not normally covered in the standard curriculum.

Part II consists of Chapters 5 through 8. Here we focus on specific aids and activities in arithmetic, algebra, geometry, and probability and statistics with a chapter devoted to each one of these four areas. Each chapter begins with some short motivational ideas and then explores several key subject matter topics in detail. Appropriate classroom activities and experiments are identified along with detailed descriptions on the construction and use of related models and aids. Extensive space is given to the use of manipulatives both in the hands of the student and in those of the teacher. They not only offer concrete models for new concepts but can be used to expand and reinforce established concepts by helping the students see familiar topics in new and different ways. Special attention is given in each chapter to the use of the overhead projector and to suggested applications using the calculator and microcomputer.

Motivation is also an underlying theme throughout Part II. Ideas are presented from the point of view of generating attention, interest, and surprise and many are tied to problem-solving situations. These are important aspects of good teaching and they can be nurtured through the effective and imaginative use of aids and activities in the classroom. This in turn enables students to see ideas presented in new and different ways and helps to foster creative thinking, perhaps the most critical of all the mathematical skills.

Much of the material that appears in this book has been presented and tested in a variety of ways. Some of the content is based on talks given by the authors at various state and national professional meetings throughout the country. Many of the ideas concerning the use of multisensory aids have been used by the authors in various undergraduate and graduate courses in mathematics education taught at Montclair State College and elsewhere. Most of the laboratory and discovery activities suggested have been used by the authors with many junior and senior high school mathematics classes that they have taught personally.

Throughout this book we shall make reference to publications of the National Council of Teachers of Mathematics, 1906 Association Drive, Reston, Virginia, 22091. A complete listing of all their publications is available free by writing to the address above.

A leading educator once said that we have done a reasonably good job in the past decade in teaching better mathematics, but that it is now time to learn to teach mathematics better. It is the fond hope of the authors that the materials and ideas presented in this book will enable teachers of mathematics to improve their pedagogical skills and thus better motivate students to learn their subject matter.

To their many students—past, present, and future—the authors fondly and encouragingly dedicate this volume as one small contribution to the big and exciting task of teaching mathematics better.

Max A. Sobel

Evan M. Maletsky

The Art of Teaching

Chapter 1

There is endless argument as to whether one is "born to teach" or can learn the skills necessary for success in this profession. Although certain basic traits seem essential, the authors believe that students and teachers with appropriate mathematical backgrounds can develop many of the requisite skills for mastering the art of teaching mathematics. Thus this first chapter lists a number of guiding principles that are essential ingredients in the artistry of teaching. Many of these principles are elaborated upon in later chapters with reference to specific subject-matter areas.

The authors once heard the job description for teachers summarized by these three statements:

Teachers must know their stuff.
They must know the pupils whom they are stuffing.
And above all, they must know how to stuff them artistically.

It is the last item of this list that we are primarily concerned with in this chapter, and indeed throughout this book. We all wish to improve our artistry in "stuffing" our students with appropriate, contemporary mathematics. This artistry is important when we attempt to motivate our students and to challenge the many reluctant learners who cross our paths daily.

Every teacher has his or her own "bag of tricks" that may work effectively. However, all conscientious teachers are constantly searching for new ideas and techniques to adopt in their classrooms. Therefore, it is hoped that the collection of items presented in this chapter may provide some additional procedures that mathematics teachers will find useful in their daily teaching.

1.1 START THE PERIOD IN AN INTERESTING WAY

There are far too many teachers of mathematics who rely on a universal lesson plan which, for a 45-minute period, proceeds as follows:

30 minutes—review yesterday's assignment
10 minutes—introduce the new lesson
 5 minutes—have students begin the next assignment

Such an approach, followed daily, can only be classified by three D's:

Dull

Deadly

Destructive of all interest

Although it may be necessary to review most assignments that are given, a teacher need not begin every lesson in this manner, and certainly need not devote the major portion of every class to such a review. The first five minutes of a period often spell the difference between the success or the failure of a lesson. Thus it is imperative that some thought be given to imaginative ways to begin so as to capture student attention and interest.

Guessing and Estimating

An interesting question can often serve as one of the most effective ways to start or end a class. A question is posed and students are given an opportunity to guess and debate the answer. Then, with teacher guidance, appropriate methods are considered for the solution of the problem. Of course the question should be designed so that its solution requires the class to employ mathematical methods appropriate to the curriculum and level of instruction at hand.

Consider, for example, a seventh-grade class studying a unit on our decimal system of notation. The teacher wishes to provide some review of fundamental computation, and also hopes to develop an appreciation of the meaning of very large numbers. An interesting question that could be used to start the class is the following:

I've just decided to count to one million.

1, 2, 3, 4, 5, . . .

How long should it take me?

Some students will begin to make random guesses. After a few of these have been offered, remind the class members that they really do not yet have adequate information for solving this problem. For example, they have not been told the rate at which you will count nor whether you will count without stopping. Tell them that you will count at the rate of one number per second and will not stop until the task has been completed. Then ask for guesses again.

Of course there will always be one bright student to give the answer as 1 million seconds! However, you should ask for the answer in more commonly understood units of time, such as days, weeks, months, or years. At this point it is extremely important to allow students to guess before they compute. A heated discussion among students concerning their guesses is the best way to motivate them to perform the computation necessary to provide the correct answer.

A word of caution should be inserted here. Occasionally one meets a class where most attempts at motivation seem to fail. After discussion of the various guesses for the time it takes to count to 1 million, you must lead the class to discover the computational method used to find the correct answer. If the class is not really interested at this point in the correct answer, there is hardly any purpose in pursuing the matter further. It is for this reason that it is important to generate sufficient discussion in advance so that students are eager to learn the solution.

This problem provides an excellent opportunity to emphasize the skills of estimation. To change 1,000,000 seconds to days we need to complete this computation:

$$\frac{1,000,000}{60 \times 60 \times 24}$$

Round 24 to 25 and note the results:

$$\frac{1,000,000}{60 \times 60 \times 25} = \frac{1,000,0\cancel{00}}{6\cancel{0} \times 6\cancel{0} \times 25} = \frac{\overset{400}{\cancel{10,000}}}{36 \times \underset{1}{\cancel{25}}} = \frac{400}{36}$$

Since $\frac{400}{36}$ is somewhat greater than 10, a very good estimate would be 11 days. (The actual answer is approximately $11\frac{1}{2}$ days.)

Challenging Questions

There are many interesting and challenging questions that can be used to stimulate discussion at the start of a period and that can also help motivate a review of computational skills. Here are just a few illustrations; others appear elsewhere in the text and within the Exercises. The reader is urged to guess and estimate for each of the following so as to experience the thought processes that students might encounter. These problems appear again in the Exercises at the end of the chapter, with answers given at the back of the book. Begin a collection of challenging questions in an index card file.

1. Consider 1 million pennies piled one on top of another. How high would the pile reach?
   As high as the ceiling of your classroom?
   As high as the school flagpole?
   As high as the Empire State Building?
   As high as the moon?

2. One million $1 bills are placed end to end on the ground. How far would they reach? Across a football field? Across the state? Across the United States? Around the world?

3. I'm going to snap my fingers. One minute later I'll snap them again. Then 2 minutes later I'll snap them again. Then I'll wait 4 minutes to snap them; then 8 minutes, 16 minutes, and so on. Each time I double the number of minutes in the interval between snaps. At this rate, how many times will I snap my fingers in one year?

4. What historical event happened approximately 1 billion seconds ago?

5. How many pencils, laid end to end, would be needed to reach from New York City to San Francisco? List the assumptions you make in estimating your answer.

As a final example, consider an eighth-grade class that is studying a unit on measurement. Here is an interesting question that could be used to start the class:

"Look at our classroom. Do you think we could fit 1 million basketballs into this room?
   How about 1 million baseballs?
   One million table-tennis balls?
   One million marbles?
   One million pennies?"

For a classroom of average size the guess is often between 1 million table-tennis balls and 1 million pennies. After time for discussion, the class should be asked to determine a procedure by which to approximate the correct answer. One possibility might be to bring an empty shoebox to class and fill it up with table-tennis balls. Then use a tape measure and approximate the volume of the room. By comparing the volume of the room with the volume of the shoebox, one can obtain a fair approximation of the number of table-tennis balls that could fit into the room.

Another question that could be used to serve the same purpose, or could be given as an assignment, is to determine the approximate weight of 1 million pennies. Questions of this type generally serve to set the stage for interesting class discussions that make students look forward to attending their mathematics class . . . a place where exciting things happen!

1.2 USE HISTORICAL TOPICS WHEN APPROPRIATE

Too many of our students think of mathematics as a very dull subject, and they picture mathematicians as hermits who spend their lives buried in mountains of figures. One interesting way to make mathematics come alive is to make frequent use of historical items that help to show that mathematicians are human beings, with mortal weaknesses and interests. Books on the history of mathematics are excellent sources for such items, many of which can be appropriately used to introduce or to supplement particular topics in the classroom.

Birthdates of Mathematicians

Make a list of birthdates of famous mathematicians and celebrate these dates when they arrive. For example, on April 30 you might consider a birthday party for the German mathematician Karl Friedrich Gauss (1777–1855), considered by many to have been one of the three greatest mathematicians of all times, along with Archimedes and Newton. Gauss has been honored by having his picture on both stamps and coins in his native country.

It is claimed that young Gauss was a precocious youngster with a born flair for mathematics. To keep him suitably occupied, his teacher in elementary school told Gauss to write the numbers from 1 to 100 and to find their sum. In a flash Gauss is supposed to have given the answer as 5050. It is further claimed that he did so by mentally recognizing this pattern:

$$1 + 2 + 3 + \cdots\cdots\cdots + 98 + 99 + 100$$

$$1 + 100 = 101$$
$$2 + 99 = 101$$
$$3 + 98 = 101$$

Since there are 50 pairs that add to 101, the total sum is 50 × 101 or 5050! Today we use the formula $S = \dfrac{n(n + 1)}{2}$ for the sum of the first n successive counting numbers.

On April 15 you might celebrate the birthday of one of the most prolific mathematicians of all times, Leonhard Euler (1707–1783). His contributions are so extensive that one can be found for almost any level of instruction. A few are noted below, along with a picture of a stamp produced in his honor in Switzerland.

Euler was the first to use the symbol π.

He was the first to use i for $\sqrt{-1}$.

The discovery of this very interesting relationship between e, π, and i is also credited to Euler:

$$e^{\pi i} + 1 = 0$$

Euler published a list of 30 pairs of *amicable numbers*. (See Exercises 15 and 16 at the end of this chapter.)

For a challenging question to start the period, ask the class why Euler was wrong when he thought that he had found a formula that would generate only prime numbers for integral values of n:

$$n^2 - n + 41$$

Another interesting contribution made by Euler was the following square array that gives a total of 260 for each horizontal row and vertical column. What is especially interesting about this array is that a chess knight can begin at the square marked 1, and can then land on each number through 64, in numerical order, by using L-shaped moves. Students who are chess players will enjoy confirming this fact, and performing "magic" tricks by memorizing the moves that produce the 64 numbers of the square.

1	48	31	50	33	16	63	18
30	51	46	3	62	19	14	35
47	2	49	32	15	34	17	64
52	29	4	45	20	61	36	13
5	44	25	56	9	40	21	60
28	53	8	41	24	57	12	37
43	6	55	26	39	10	59	22
54	27	42	7	58	23	38	11

Any of these contributions by Euler, or others found in history of mathematics books, can serve to generate student interest in various topics under discussion in the classroom.

Here are the birthdates of some famous mathematicians that can be used to motivate a discussion of their lives and work. The list is only a representative one and can be enlarged by referring to history of mathematics books. (Also see the calendar published by the National Council of Teachers of Mathematics.)

Jan. 3	Sonya Kovalevsky (1850–1891)
Jan. 23	Davis Hilbert (1862–1943)
Feb. 19	Nicolaus Copernicus (1473–1543)
Mar. 3	Georg Cantor (1845–1918)
Mar. 14	Albert Einstein (1879–1955)
Mar. 23	Emmy Noether (1882–1935)
Apr. 1	Sophie Germain (1776–1831)
Apr. 15	Leonhard Euler (1707–1783)
Apr. 30	Karl Friedrich Gauss (1777–1855)
May 16	Maria Gaetana Agnesi (1718–1799)
May 26	Abraham De Moivre (1667–1754)
June 19	Blaise Pascal (1623–1662)
July 1	Gottfried Leibniz (1646–1716)
Aug. 5	Neils Henrick Abel (1802–1829)
Aug. 20	Pierre de Fermat (1601–1665)
Sept. 17	Bernhard Reimann (1826–1866)
Oct. 25	Evariste Galois (1811–1832)
Nov. 17	August Möbius (1790–1868)
Dec. 2	Nicolai Lobachevsky (1792–1856)
Dec. 25	Isaac Newton (1642–1727)

A similar listing for major contributions in the development of computers can be of interest to students. Including two from the preceding list, the following individuals played key roles in bringing about our current age of technology. Suggest that your students find the major contributions made by each.

Howard Aiken (1900–1973)
Charles Babbage (1791–1871)
J. Presper Eckert (1919–   )
Herman Hollerith (1860–1929)
Joseph Jacquard (1752–1834)
Gottfried Leibniz (1646–1716)
Augusta Ada Byron Lovelace (1815–1852)
John Mauchly (1907–1980)
John Napier (1550–1617)
John Von Neumann (1903–1957)
Blaise Pascal (1623–1662)

Quotable Quotes

Quotations from famous mathematicians make effective bulletin board displays. Here are several noteworthy ones:

"Mathematics is the queen of the sciences, and arithmetic is the queen of mathematics." (Karl Friedrich Gauss)
"God created the natural numbers; everything else is man's handi-work." (Leopold Kronecker)
"Give me a place to stand and a lever long enough and I will move the earth." (Archimedes)

In 387 B.C., the great Greek philosopher Plato founded his famous Academy in Athens for the pursuit of philosophical and scientific inquiry. Over its door was the motto:

"Let no one ignorant of geometry enter here."

Unsolved Problems

Unsolved and impossible problems are usually of great interest to mathematics students. Every teacher of geometry has had experience with would-be angle trisectors, students who attempt the impossible. When studying the Pythagorean theorem, point out that $x^2 + y^2 = z^2$ has many integer solutions for x, y, and z, such as 3, 4, 5 and 5, 12, 13. Then tell this interesting anecdote about an unsolved problem known as Fermat's Last Theorem.

Pierre de Fermat (1601–1665) claimed that there do not exist positive integers, x, y, and z such that $x^n + y^n = z^n$, for $n > 2$. Fermat claimed that he had discovered a proof of the impossibility of this relation but that the margin of his book was too narrow to include it. To this date no mathematician has been able to prove the impossibility of this problem. It is therefore felt that Fermat believed he had a proof, but that there was undoubtedly some flaw in it; of course, that is just a hypothesis.

In the early 1900s, a professor of mathematics in Darmstadt, Germany, Paul Wolfskehl, spent many long hours in a vain attempt to prove Fermat's Last Theorem. He was also disappointed in love, and so decided to end his life. Being a methodical man, he wrote a suicide note specifying the date and hour that he would commit the act. Within a few hours of the appointed time he decided to occupy his final hours by one last look at Fermat's Theorem. It is said that he became so engrossed once again in the problem that his appointed suicide hour passed by unnoticed, whereupon he tore up his note and began life with renewed energy. When Wolfskehl finally died in 1908 he left a will that provided 100,000 marks to the first one to prove Fermat's Theorem, a sum long since depleted by inflation.

There are many ways that the creative teacher can use this story to generate interest in a classroom. Just start with a simulated newspaper headline on the chalkboard or bulletin board that states

MATHEMATICS SAVES A LIFE!

The story of Paul Wolfskehl can be an exciting introduction to the study of the Pythagorean theorem.

Historical Anecdotes

The history of mathematics is rich with interesting anecdotes about mathematicians that can be used to generate interest in a topic under study in the classroom. Here are a few more examples.

Maria Gaetana Agnesi (1718–1799) was a somnambulist who often solved problems in her sleep and awoke to find the completed solutions on her desk in the morning! She is probably best known for a curve, now known as the "Witch of Agnesi," that she discussed in detail. Secondary students can be asked to sketch this curve.

$$y = \frac{a^3}{a^2 + x^2}$$

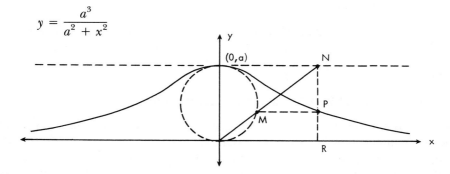

The curve is the locus of points P that may be found geometrically as follows. Draw a circle with diameter of length a that passes through the origin and the point $(0, a)$. Draw a secant from the origin that meets the circle at point M and meets the line through $(0, a)$, parallel to the x-axis, at N. Draw the segment NR perpendicular to the x-axis. Then draw a line segment through point M, parallel to the *x*-axis, to locate the point P.

Lorenzo Mascheroni (1750–1800) proved that all Euclidean construction can be completed with a compass alone. (It is understood that two points determine a unique line, but that the line itself cannot be drawn.) This mathematician met Napoleon during his conquest of Italy and was challenged to show how a circle can be divided into four equal parts by compass alone. Both the story and the problem are excellent items to incorporate into a lesson for a geometry class studying construction.

Surprisingly, most secondary students of mathematics appear to be quite interested in historical items. Ancient systems of computation, for example, can be used quite effectively to enliven a class and serve well as an introduction to modern approaches. Two specific examples, one arithmetic and the other algebraic, are given here.

The ancient Egyptian method of duplation and mediation for multiplication fascinates most students. To multiply two numbers, such as 25×32, one doubles one of the factors while halving the other. When dividing by 2, all remainders are discarded, as shown below. Next, all the odd numbers in the first column are circled. The numbers opposite these in the second column (32, 256, 512) are added. Their sum, 800, represents the product 25×32.

DIVIDE BY 2	MULTIPLY BY 2	
(25)	32 ✓	
12	64	Note that 25 ÷ 2 is written as 12, and
6	128	3 ÷ 2 is written as 1 since remainders
(3)	256 ✓	are discarded.
(1)	512 ✓	

Answer: $25 \times 32 = 32 + 256 + 512 = 800$

Junior high school students enjoy trying this method for various products, and it serves well as a means of motivating computation. A discussion of binary notation can be used to justify the procedure.

1.3 MAKE EFFECTIVE USE OF MANIPULATIVE AIDS

The chapters that follow provide numerous suggestions for aids that can be used in the mathematics classroom to promote learning. The major emphasis throughout is on aids that the teacher can produce with a minimum of time and effort as opposed to more costly commercial products. Although many of the latter are worthwhile, most teachers of mathematics are more receptive to the use of aids if they can be quickly assembled and used.

Using Paper as an Aid

What uses can be found for a strip of paper off a roll of register tape? Some teachers would use it to bring a **googol** to class. A googol can be written as 1 followed by 100 zeros, or simply as 10^{100}. The number doesn't sound too big when you talk about it, but students can begin to appreciate just how large it is when the teacher pulls out a strip of paper and unrolls the 100 zeros clear across the classroom. By comparison, 1 million looks trivial.

GOOGOL 10,000,000,000,000,000,000,000,000,000,000,
               000,000,000,000,000,000,000,000,000,000,000,
               000,000,000,000,000,000,000,000,000,000,000

Patterns for representing infinite repeating and nonrepeating decimals can also be effectively displayed this way on strips of paper.

repeating   .311311311311311311311311311311311311311311 . . .

nonrepeating .311331133331133333113333331133333331133333331 . . .

When discussing the nonrepeating nature of the decimal representation for the irrational number π, many students nod in agreement but do not really believe that it *never* repeats. The message is much more impressive and lasting when the students can actually see and search 100 or so of the digits for repeating patterns. All the teacher needs is a strip of paper and the time to copy the digits down.

π = 3.14159 26535 89793 23846 26433 83279 50288 41971
           69399 37510 58209 74944 59230 78164 06286 20899
           86280 34825 34211 70679 . . .

A listing of the first 4000 decimals in the expansion of π can be found in *The Lore of Large Numbers*, by Philip J. Davis (Washington, D.C.: The Mathematical Association of America, 1961, pp. 72–73).

Paper-Folding and Cutting Activities

A sheet of paper can also be a simple, handy aid for the mathematics teacher. Paper-folding activities frequently stimulate interest in geometry as well as provide challenging problems. For example, when studying equilateral triangles, students can explore methods for constructing one by folding paper using the shorter edge as one side:

Start with one
edge as a side.

Fold the
perpendicular
bisector of
that edge.

Locate the
vertex on
that
bisector.

Fold through that
point to complete
the equilateral
triangle.

Another activity with paper might involve an informal proof, as in this illustration of the Pythagorean theorem:

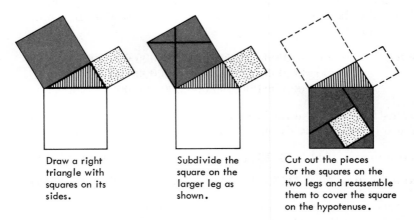

| Draw a right triangle with squares on its sides. | Subdivide the square on the larger leg as shown. | Cut out the pieces for the squares on the two legs and reassemble them to cover the square on the hypotenuse. |

At a more advanced level, the teacher can introduce the topic of finding the sum of an infinite geometric series by duplicating an experiment that is said to have been carried out by Archimedes. Begin with a square sheet of paper and consider this as having an area of one square unit. Cut the paper in half, and place one half on the desk. Cut the remaining piece in half so as to obtain two pieces, each with an area of 1/4 square unit. Place one of these pieces on the desk. Continue cutting in this manner until the piece that remains in your hand is too small to handle, but imagine the process going on forever. Then note that the sum of the areas of the pieces represent

$$\frac{1}{2} + \frac{1}{4} + \frac{1}{8} + \frac{1}{16} + \frac{1}{32} + \frac{1}{64} + \frac{1}{128} + \dots$$

which, in the limit, is equal to the original area of one square unit. This provides the necessary motivation to introduce the formula for the sum of any such infinite geometric series where $|r| < 1$:

$$S = \frac{a}{1 - r}$$

For the preceding series,

$$a = 1/2, \quad r = 1/2, \quad \text{and } S = \frac{\frac{1}{2}}{1 - \frac{1}{2}} = 1$$

The reader should attempt to expand this activity by dividing the original unit square into thirds. Place two of these pieces on the desk and cut the remaining piece into thirds again. Continue in this manner so that each of the two piles has an area equal to the sum

$$\frac{1}{3} + \frac{1}{9} + \frac{1}{27} + \frac{1}{81} + \cdots$$

Since there are two equal piles, each must have one half of the original area of 1. Hence, the sum shown must equal $\frac{1}{2}$. Now try to generalize this procedure by considering divisions of the original piece of paper into fourths, fifths, and so on.

Using a Piece of String

What uses can the teacher find for a piece of string in the mathematics class? Come to class with a piece stuffed in your pocket and start pulling it out. Ask each student to guess at the length before he or she sees it all. Obviously, the answers will vary widely. As the rest is pulled out for them to see, students again guess at the length. The answers will still vary, probably more than the students themselves would expect. Next they are asked to guess how many times it will fit around a dollar bill and a basketball. Once the actual length is given, the ends are tied together and students are asked for the dimensions of the largest square and equilateral triangle that can be formed from it. As a last activity, the teacher can go from one student to the next throughout the entire class asking for a different set of dimensions for a rectangle or an isosceles triangle that can be formed with the string. With a simple piece of string, students can become actively involved and motivated to a further study of measurement. (Also see page 255.)

1.4   MAKE PROVISIONS FOR STUDENT DISCOVERY

Conflicting points of view exist concerning the role of discovery in the teaching of mathematics. Some claim that skills and concepts are best learned when students are allowed to make significant discoveries on their own. On the other hand, others feel that many students, especially slow learners, learn best by means of a teacher-oriented show-and-tell approach. Regardless of which position one takes, it is clear that discovery techniques can be used effectively to stimulate and maintain interest in mathematics. Furthermore, such approaches help to develop the type of creativity and originality that is important for a student's future success in mathematics.

The famous French mathematician René Descartes (1596–1650) concluded his book *La Geometrie* with this comment:

> I hope that posterity will judge me kindly, not only as to the things which I have explained, but also as to those which I have intentionally omitted so as to leave to others the pleasure of discovery.

In our desire to impart knowledge to our students, we must not fail to make provisions for them to participate in and to enjoy this all-important pleasure.

Actually there are two different types of discovery approaches that can be used in the classroom: *guided discovery* and *creative discovery*.

Guided Discovery

Most classroom situations lend themselves best to the guided discovery approach, in which the teacher leads a class along the right path, rejecting incorrect attempts, asking leading questions, and introducing key ideas as necessary. It is a cooperative venture that becomes more and more exciting as the final result comes into view.

The following items should serve to illustrate guided discovery approaches that can be utilized in the classroom with most students.

1. A class is asked to find this sum:

$$\frac{1}{1\cdot2} + \frac{1}{2\cdot3} + \frac{1}{3\cdot4} + \cdots + \frac{1}{99\cdot100}$$

The task appears to be impossible. The teacher suggests that one problem-solving strategy is to consider a small part of the problem at a time. Thus the class is led to consider the first term, the first two terms, the first three terms, and so on.

$$\frac{1}{1\cdot2} = \frac{1}{2}$$

$$\frac{1}{1\cdot2} + \frac{1}{2\cdot3} = \frac{2}{3}$$

$$\frac{1}{1\cdot2} + \frac{1}{2\cdot3} + \frac{1}{3\cdot4} = \frac{3}{4}$$

At this point the teacher asks the class to guess the sum of the first four terms, pointing out the pattern if necessary. Hopefully there will be members of the class who will guess that it will be $\frac{4}{5}$. This answer is confirmed by actual computation. Finally, the class should be ready to guess that the answer to the given problem is $\frac{99}{100}$. Of course, it is important to point out that this is just a conjecture and not a proof.

We can prove this sum of $\frac{99}{100}$ by elementary methods. First we need to recognize these relationships, which also lend themselves to discovery approaches.

$$\frac{1}{1\cdot2} = \frac{1}{2} = \frac{1}{1} - \frac{1}{2} \qquad \frac{1}{2\cdot3} = \frac{1}{6} = \frac{1}{2} - \frac{1}{3} \qquad \frac{1}{3\cdot4} = \frac{1}{12} = \frac{1}{3} - \frac{1}{4}$$

Then write the given series as follows:

$$\left(1 - \frac{1}{2}\right) + \left(\frac{1}{2} - \frac{1}{3}\right) + \left(\frac{1}{3} - \frac{1}{4}\right) + \cdots + \left(\frac{1}{98} - \frac{1}{99}\right) + \left(\frac{1}{99} - \frac{1}{100}\right)$$

Finally, note that every term except the first and last subtract out, giving this sum:

$$1 - \frac{1}{100} = \frac{99}{100}$$

2. As has already been stated, it is important that students recognize that a conjecture is not a proof, and that without a proof there is no guarantee that a pattern will continue forever. Thus it is worthwhile to occasionally display a pattern that fails after a certain point. One of the most dramatic ones concerns the maximum number of regions into which a circle can be divided by connecting points on the circle. Consider the following apparent pattern:

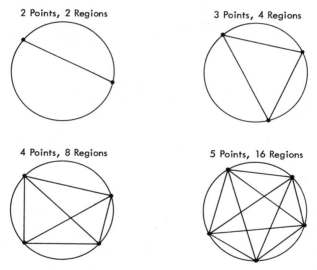

2 Points, 2 Regions

3 Points, 4 Regions

4 Points, 8 Regions

5 Points, 16 Regions

In summary:

NUMBER OF POINTS	NUMBER OF REGIONS
2	2
3	4
4	8
5	16

What is your conjecture about the number of regions expected for six points connected in all possible ways?

Test your conjecture on another circle. The apparent answer is 32. However, much to everyone's surprise, the maximum number of regions possible with six points proves to be only 31!

3. Assume you earn $1 on the first day of a job. The second day you are paid $2, the third day $4, the fourth day $8, and so on. Each day your salary is twice that of the preceding day. You plan to stay on the job for 15 days and wish to know what your total earnings will be. The specific amounts for each day could be listed, and their sum found, but you ask the class to discover a method for finding the sum without actually adding the 15 amounts.

An important problem-solving strategy is to consider a small portion of a problem at first (see Chapter 3). That is, consider what the total salary will be for 1 day, 2 days, 3 days, 4 days, and then for 5 days.

FOR 1 DAY	FOR 2 DAYS	FOR 3 DAYS	FOR 4 DAYS	FOR 5 DAYS
1	1	1	1	1
Total 1	2	2	2	2
	Total 3	4	4	4
		Total 7	8	8
			Total 15	16
				Total 31

At this point the class is asked to observe relationships. With teacher assistance they should note these facts:

The salary for 2 days ($3) is $1 less than the salary for the third day ($4).

The salary for 3 days ($7) is $1 less than the salary for the fourth day ($8).

The salary for 4 days ($15) is $1 less than the salary for the fifth day ($16).

The salary for 5 days ($31) is $1 less than the salary for the sixth day ($32).

The discovery is made that the total salary for n days is 1 less than that for the $(n + 1)$ day. To find the salary for 15 days, add

$$1 + 2 + 4 + 8 + 16 + 32 + 64 + 128 + 256 + 512$$
$$+ 1024 + 2048 + 4096 + 8192 + 16,384$$

This sum is 1 less than the next term: $2 \times 16,384 - 1 = 32,767$. Thus the total salary for 15 days is $32,767.

At another level of instruction, students might note that we are dealing here with powers of 2:

Day 1	1	2^0	
Day 2	2	2^1	
Day 3	4	2^2	The sum of the entries for
Day 4	8	2^3	the first n days is $2^n - 1$.
.			
.			
.			
Day n	.....	2^{n-1}	
Day $n + 1$	.....	2^n	

Creative Discovery

Many mathematics textbooks claim to utilize a discovery approach. However, it is difficult to do so in a text where final results must be stated. Thus most discovery techniques must come about through the direction of the classroom teacher. However, at its purest level we have creative discovery, wherein a teacher presents a situation to a class and allows the students to explore on their own, using only their intuition and past learning, with little or no guided direction. Such an approach is especially well suited to the gifted student and provides the type of experience that is necessary for later independent research.

Essentially, the creative discovery approach says

Here's a situation . . . explore it.

Most students are disturbed by such an instruction because it is foreign to the traditional problem-solving techniques they have encountered in the past. For example, students might be given this triangular array of numbers:

$$1$$

$$3 \quad 5$$

$$7 \quad 9 \quad 11$$

$$13 \quad 15 \quad 17 \quad 19$$

. . . .

They are then asked to make discoveries about it, without any further direction. Among the many discoveries that might be made is the fact that the sum of the numbers in each successive row represents the cubes of the counting numbers: 1, 8, 27, 64, . . . Can you make some other discoveries?

As another example, consider this table of counting numbers:

1	2	3	4	5	6	7	8	9	10
11	12	13	14	15	16	17	18	19	20
21	22	23	24	25	26	27	28	29	30
31	32	33	34	35	36	37	38	39	40
41	42	43	44	45	46	47	48	49	50
51	52	53	54	55	56	57	58	59	60
61	62	63	64	65	66	67	68	69	70
71	72	73	74	75	76	77	78	79	80
81	82	83	84	85	86	87	88	89	90
91	92	93	94	95	96	97	98	99	100

Students are given the table and asked to see what discoveries they can make. A few of the possibilities are given below; the reader will be asked to search for others in the Exercises.

1. The multiples of 9 fall on a diagonal line, as do the multiples of 11.
2. For any rectangular array of numbers within the table, the sums of the entries in opposite corners are equal. That is,

$$
\begin{array}{cccc}
⑦② & 73 & 74 & ⑦⑤ \\
82 & 83 & 84 & 85 \\
⑨② & 93 & 94 & ⑨⑤
\end{array}
$$

$$72 + 95 = 167 = 75 + 92$$

3. For any square array of nine numbers, the sum of all entries is equal to nine times the number in the center of the square.

$$
\begin{array}{ccc}
47 & 48 & 49 \\
57 & ⑤⑧ & 59 \\
67 & 68 & 69
\end{array}
$$

$$47 + 48 + 49 + 57 + 58 + 59 + 67 + 68 + 69 = 522 = 9 \times 58$$

A more sophisticated example of the creative discovery approach can be given to an eleventh- or twelfth-grade class that has already studied a unit of work on arithmetic and geometric series. The teacher proceeds to tell the class about the infinite tree. This tree looks like any other tree but grows in a very interesting manner. The first day the tree grows 1 foot. The second day two new branches grow, each $\frac{1}{2}$ foot in length and at right angles to each other. The next day two new branches appear at each terminal point, again at right angles to one another but only $\frac{1}{4}$ foot in length. This continues forever! This is what the tree looks like during the first four days of growth.

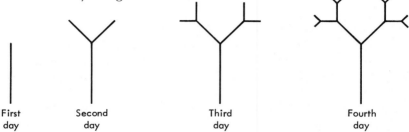

First day   Second day   Third day   Fourth day

At this point in the typical mathematics class the teacher would ask the students to prove certain given relationships. But using a creative discovery approach, the teacher terminates the story of the infinite tree by asking the students to discover whatever they can about it, offering no further clues or direction. Among the many interesting items that students may discover, using only right-triangle relationships and knowledge of geometric series, is that the tree has a limiting finite height of $\dfrac{4 + \sqrt{2}}{3}$ feet and a limiting breadth of $\dfrac{2(\sqrt{2} + 1)}{3}$ feet, even though there are an infinite number of branches with a total length that is also infinite. (See Exercise 12 and Activity 18 at the end of this chapter.)

A similar problem is that of the infinite snowflake. It hits the ground in the shape of an equilateral triangle. Thereafter, each second a new equilateral triangle emerges in the middle third of each side, continuing forever. This is what the first three stages look like:

Once again the student is asked to discover whatever he or she can about the infinite snowflake, with no teacher direction. Among the many interesting facts that can be discovered is that there is no limiting perimeter to this curve but that there is a limiting area, $\dfrac{2}{5} n^2 \sqrt{3}$ square units, where n is the length of the side of the original equilateral triangle. (See Exercise 13 at the end of this chapter.)

In closing, it should be noted that the creative discovery approach is especially well suited to gifted students. The average or slow learner can seldom function without any teacher direction at all. For the latter group, it is important that we do not deny them the pleasures of discovery that can be attained through teacher direction.

1.5 END THE PERIOD WITH SOMETHING SPECTACULAR

This bit of advice is not always easy to follow inasmuch as the supply of truly spectacular items is limited. However, through the years a teacher should be able to collect enough interesting material to avoid ending classes in the traditional way: "Begin your homework." With less than five minutes left in the period, this suggestion is seldom effective nor met with enthusiasm!

Introducing some special item during the last few minutes of a class can make students regret that it has come to a close. Hopefully they will walk out talking about the exciting things that happen in their mathematics period . . . and hopefully their enthusiasm will carry over to the next day when they will be eager to return for more.

Magic Numbers

Many spectacular things can be done by embellishing rather simple mathematical tricks or patterns. Several examples are given below, others are inserted throughout the text, and still others may be found by a careful search of the available literature.

1. A student is asked to go to the board. With the teacher's back turned so as not to see the work, the student is given these instructions:

> Write a two-digit number between 50 and 100.
> Add 76 to this number.
> Cross out the digit in the hundreds place.
> Add the crossed out number to the remaining two-digit number.
> Subtract this result from the original number.

These are the steps if a student were to begin with the number 83:

$$
\begin{array}{lr}
\text{Original number:} & 83 \\
\text{Add 76:} & +\ 76 \\
\hline
& 159
\end{array}
$$

$$
\text{Cross out and add:} \quad \cancel{1}59; \quad 59 + 1 = 60
$$

$$
\begin{array}{lr}
& 83 \\
\text{Subtract from original number:} & -\ 60 \\
\hline
\text{Result:} & 23
\end{array}
$$

The interesting thing about this trick is that the final outcome will always be 23, regardless of the number selected by the student, provided that the stated steps are followed. However, it is certainly not very spectacular to conclude by announcing that the final outcome is 23, as interesting as this may be. A far more dramatic approach is the following.

Before coming to class use the edge of a damp piece of soap and write 23 on the back of your hand. When this dries it becomes completely invisible to the student. In class, after completing the puzzle, ask a member of the class to write the final outcome on a piece of paper and fold it. Then carefully burn the paper in some suitable receptacle and wait until the ashes cool off. Finally pick up the ashes and wipe them on the back of your hand, whereupon, as if by magic, the number 23 is clearly outlined for all to see! It has been the experience of the authors that this trick is one that students continue to talk about for many weeks.

As an alternative ending to this trick, use lemon juice and a toothpick to write 23 on a clear sheet of acetate. This will be invisible when placed on an overhead projector. However when ashes are rubbed over the acetate, the number 23 will magically appear!

2. Have each student in the class write a four-digit number, using four different digits. Then form three additional *cyclic numbers* by moving the digit in the thousands place to the hundreds, tens, and units place. Here is an example, using as the initial number 8234:

$$8234$$
$$4823$$
$$3482$$
$$2348$$

Have students find the sum of their four numbers.

(For the example above, the sum is 18,887.)

Find the sum of the digits in the original number.

$$(8 + 2 + 3 + 4 = 17)$$

Divide the sum of the numbers by the sum of the digits.

$$(18,887 \div 17 = 1111)$$

Surprisingly, regardless of the original number chosen, the sum will always be 1111!

Mathematical Fallacies

Fallacies are generally of interest to secondary mathematics students. An interesting way to end a class period is to announce the recent discovery of a proof that 1 = 2! This is a standard fallacy that appears in many algebra texts.

Let $a = b$
Then $a \cdot a = a \cdot b$; that is, $a^2 = ab$
By subtraction, $a^2 - b^2 = ab - b^2$
By factoring, $(a - b)(a + b) = b(a - b)$
By division, $a + b = b$
Thus, $b + b = b$ since $a = b$
Finally, $2b = b$ and $2 = 1$

Of course, the fallacy lies in the fact that we divided by 0 in the form of $a - b$. This can then lead to a well-motivated discussion of why division by zero is not permissible.

Problem of the Week

An interesting way to end a period is with a Problem of the Day or Problem of the Week. This should be some challenging problem or puzzle with a solution that is not immediately obvious. Students then are given a day or a week to solve it. Reward the first person to submit a correct solution, and provide honorable mention to everyone who submits a solution. Here is one problem that has proven to be quite challenging to students of all ages:

Find the pattern:

$$3 * 4 \rightarrow 5$$
$$4 * 7 \rightarrow 1$$
$$8 * 4 \rightarrow 0$$
$$1 * 2 \rightarrow 9$$

The challenge is to discover how the third number is obtained from the first two numbers. That is, if you know the pattern, then you will be able to complete these statements:

$$5 * 5 \rightarrow ? \qquad 4 * 1 \rightarrow ? \qquad 6 * 2 \rightarrow ?$$

As a tantalizing hint, note that this problem first appeared in a third-grade text! (What computational skills do third graders possess? See Exercise 17.)

Here are several other problems that can be used as puzzles of the day or of the week. (See Exercises 18 and 19.)

1. Discover the pattern and find the next two numbers in each of these sequences:

$$12, 1, 1, 1, 2, 1, \ldots$$
$$0, 6, 20, 42, \ldots$$

2. Find the pattern for generating the figures in this chart:

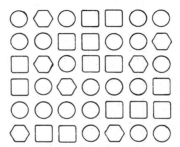

1.6 CONCLUSION

As originally anticipated, a chapter on "The Art of Teaching" is in reality both an impossible and an endless task at the same time! The list of topics that one might include is long and yet can never be all-inclusive. Indeed, an excellent exercise for the reader is to attempt to constantly add to this initial list of items.

Throughout this discussion the personal qualities of the teacher have been bypassed. The easiest way to consider this item is to ask a group of secondary school students to submit an unsigned list of what they consider to be the five most important qualities of a good teacher. The results are predictable. High on the list will be such items as enthusiasm, sincerity, sense of humor, empathy, imagination, and competence. In addition, in their own words, they will say that they like teachers who like to teach.

Thus by our every action we must show our students that we like to teach mathematics. Enthusiasm is contagious, and when teachers demonstrate a sincere interest in both their students and their subject, their classes will seldom ask the embarrassing question, "What good is all this?" The authors of this book sincerely hope that the material in the chapters to follow will help the teacher of mathematics to teach in an exciting, meaningful, and memorable manner.

EXERCISES

(Where appropriate in the following Exercises, first guess or estimate the answer.)

1. One million pennies are piled one on top of another. How high will the pile be?

2. One million $1 bills are placed end to end on the ground. Approximately how far will the bills stretch?

3. Snap your fingers. One minute later snap them again. Then wait 2 minutes before the next snap and 4 minutes for the following snap. Then wait 8 minutes. Continue in this manner, each time doubling the interval between snaps. At this rate, how many times will you snap your fingers in one year?

4. In Exercise 3, change the intervals to seconds instead of minutes. How many times will you snap your fingers in one year?

5. How many pencils, laid end to end, are needed to reach from New York City to San Francisco? (What assumptions must be made?)

6. What historical event happened approximately 1 billion seconds ago?

7. Use Gauss' method, described on page 5, to find the sum of the first n counting numbers.

8. Find a value for n for which the formula $n^2 - n + 41$ does *not* produce a prime number.

9. Use the Egyptian method of duplation and mediation to find the product 35×61.

10. For the table of counting numbers on page 18, prove algebraically that the products of the entries in opposite corners of a square array of nine numbers will always differ by 40.

11. In Exercise 10, consider square arrays of 16 numbers. Show that the difference of the products of the entries in opposite corners is constant. What is the constant?

12. For the infinite tree described on page 18 show that the limiting height is $\dfrac{4 + \sqrt{2}}{3}$ feet, and the limiting breadth is $\dfrac{2(\sqrt{2} + 1)}{3}$ feet.

13. For the infinite snowflake described on page 19 show that there is no limiting perimeter but that the limiting area is $\dfrac{2}{5} n^2 \sqrt{3}$ square units, where n is the length of a side of the original triangle.

14. Explain why the "23" trick described on page 20 works.

15. Two numbers are said to be amicable if the sum of the proper divisors of one number is equal to the other number. (The proper divisors of a number are all the factors of that number except the number itself. For example, the proper divisors of 6 are 1, 2, and 3.) Show that 220 and 284 are amicable numbers.

16. One of the pairs of amicable numbers that Euler omitted was found in 1866 by a 16-year-old boy in Italy, Nicolo Paganini. Show that 1184 and 1210 are amicable numbers.

17. Find the pattern for the pattern of numbers given on page 22.

18. Find the pattern for the sequences given on page 22 and find the next two numbers in each.

19. Find the pattern in the chart of figures given on page 22.

20. Given points A and B, use a compass only to find a point C that is collinear with A and B and such that AC = 2(AB).

ACTIVITIES

1. Prepare a set of five questions that can be used in a junior high school mathematics class to promote guessing, and subsequent discovery of an answer. If possible, test these items in an actual class situation.

2. Repeat Activity 1 for a senior high school mathematics class.

3. Repeat Activity 1 for a ninth-grade general mathematics class of slow learners.

4. Prepare a collection of historical anecdotes about mathematicians that could be used in a mathematics classroom to motivate students.

5. Develop a one-period lesson plan for a junior high school mathematics class that features a discovery approach.

6. Repeat Activity 5 for a senior high school mathematics class.

7. Compile a list of different ways that a piece of paper can be used in the mathematics classroom as a visual aid. Do not neglect graph paper.

8. Ask a class of secondary school mathematics students to prepare a list of what they consider to be the most important qualities of a good teacher. Then summarize the results in order of the most frequently mentioned items.

9. Prepare a collection of at least 15 problems or puzzles that could be used effectively to start a mathematics period in an interesting way.

10. Prepare a collection of at least 15 challenging problems that could be used as the Problem of the Week in a junior high school mathematics class.

11. Repeat Activity 10 for a senior high school mathematics class.

12. Prepare a collection of algebraic and geometric fallacies and present one of these to your class.

13. List at least five different ways to review a homework assignment.

14. Prepare an ordinary multiplication table for the facts through 10×10. Then see how many different discoveries you can make concerning the entries within the table.

15. For the table in Activity 14, find the sum of all the entries without adding. (*Hint:* Use the strategy of exploring a simpler, related problem first. Thus consider a table of facts through 2×2, 3×3, 4×4, etc.)

16. Extend the collection of birthdates of mathematicians given on page 7.

17. Begin a collection of quotations from famous mathematicians.

18. LOGO allows procedures to be written that call themselves. This idea of recursion is illustrated here in this graphics program for the infinite tree problem given on page 000. Run the program using an input value of 50. Then explain how the procedure uses recursion to draw the tree.

```
TO TREE :B
HT
FULLSCREEN
REPEAT 2[RT 180 FD 2*:B]
IF :B < 1[STOP]
RT 45 FD :B
TREE :B/2
BK :B
LT 90 FD :B
TREE :B/2
BK :B RT 45
END
```

19. Use the idea of recursion to write a LOGO program that will draw the first five stages of the infinite snowflake problem described on page 19.

READINGS AND REFERENCES

1. Read about the Snowflake Curve in Chapter 9 in *Mathematics and the Imagination* by Edward Kasner and James Newman, New York: Simon and Schuster, 1967. Report on what the authors call the Anti-Snowflake Curve, the In-And-Out Curve, and Space-Filling Curves, with suitable illustrations for each type.

2. Read the tragic story of Evariste Galois in *Whom The Gods Love* by Leopold Infeld, a publication of the National Council of Teachers of Mathematics in their "Classics in Mathematics Education" series (1978).

3. Prepare a report on the lives of three of the women mathematicians discussed

in *Math Equals* by Teri Perl, Menlo Park, CA: Addison-Wesley Publishing Company, Inc., 1978.

4. Read the story of Fermat in Chapter 4 in *Men of Mathematics* by E.T. Bell, New York: Simon and Schuster, 1965. Explain why the chapter is entitled "The Prince of Amateurs."

5. The February 1983 issue of the *Mathematics Teacher*, a publication of the National Council of Teachers of Mathematics, contains an article by Lowell Leake entitled "What Every Secondary School Mathematics Teacher Should Read—Twenty-Four Opinions." Study this article and report on the recommended reference texts in the area of enrichment of the curriculum.

6. Prepare a report on Mascheroni constructions. Among the references to consult is *What is Mathematics?* by Richard Courant and Herbert Robbins, New York: Oxford University Press, 1941. Then attempt to divide a circle into four equal parts by compass alone.

7. Read *Activities for the Maintenance of Computational Skills and the Discovery of Patterns*, a 1980 publication of the National Council of Teachers of Mathematics by Bonnie H. Litwiller and David R. Duncan. Prepare a report on several of the patterns discussed by the author.

8. Prepare a report on prime numbers and their history. In particular, report on Capsules 14 and 15 found in the 31st Yearbook of the National Council of Teachers of Mathematics titled *Historical Topics for the Mathematics Classroom*, 1969.

9. Prepare a report on ancient methods of computation, referring to a history of mathematics text such as *An Introduction to the History of Mathematics* by Howard Eves, New York: Holt, Rinehart, and Winston, Inc., 1969.

10. Read the two volumes titled *Mathematical Discovery* by George Polya, New York: John Wiley & Sons, Inc., 1962, 1965. In particular, report on Chapter 13 in Volume II, "Rules of Discovery?" Also read Chapter 14 of that volume and report on Polya's "Ten Commandments For Teachers."

11. Read Chapter 2, "History of the Computer" in *Computers in Today's World* by Gary G. Bitter, New York: John Wiley & Sons, 1984.

Motivating
Mathematical Learning

Chapter 2

Almost every mathematics teacher will agree on the importance of proper motivation for the teaching of mathematics. Students, except for the very few who seem to have a natural love for the subject, need to have their interest stimulated through suitable teaching techniques and procedures. Only by doing so can we avoid problems such as math anxiety, which has become so prevalent in recent years.

Students work most effectively if they are truly interested in the subject at hand. However, it is difficult for teachers to locate a supply of interesting materials and ideas. Many teachers become so involved with the routines of presenting their subject matter that they lack the necessary time and energy to search for motivational items. Nevertheless, there is an abundant supply, and this chapter presents a small sampling that will hopefully encourage the reader to search for similar ideas.

The suggestions that follow in this chapter and in subsequent chapters may be used in the classroom in many different ways. Some are helpful to introduce a specific topic, whereas others are designed to show applications of a unit already studied. Most of them, however, are intended for use during the first or last few minutes of a period—to start the class off with a bang and/or to maintain students' attention during those last five minutes before the bell. They are all designed to help develop the idea that mathematics can be both interesting and fun. Hopefully, their use will make students look forward to the mathematics class and feel sorry to see the period end!

2.1 PROVIDE OPPORTUNITIES FOR GUESSING AND ESTIMATING

George Polya of Stanford University has said that "mathematics in the making consists of guesses." In order to make a discovery, it is first necessary to make a guess . . . and the guess may be hasty; indeed it should be. These guesses then need to be followed by verification, and this is the hard part of mathematics, the proof in support of a guess. But proof is also the least imaginative part of the process; it is the original guess that is the creative part of mathematics.

Most adults are afraid to take a guess for fear of being wrong. On the other hand, most adolescents are ready and eager to guess and the teacher should capitalize on this by providing suitable opportunities for intuitive guessing to take place in the classroom.

It is important that students be given sufficient time to formulate guesses and discuss these in class before attempting to find a correct answer through computation. Unless time is taken for this intuitive discussion, the topic serves only to provide a vehicle for computation, and the motivational aspects are lost. Consider, for example, the following problem that could be posed at the start or end of a period in a junior high school class:

> Look at this piece of paper that I am holding.
> I'm going to fold it once so as to have 2 pieces.
> Now I'll fold it again so as to have 4 pieces.
> I'll continue folding so as to have 8 pieces, 16 pieces, and so on.
> Assume that I'm able to continue in this way for 50 folds. How thick would the stack of paper be?

As indicated in Chapter 1, if the teacher immediately turns to the task of finding an answer by computation, the problem loses all interest to the student. Rather, one should urge students to guess and allow the class to establish the range between the least and the greatest answers given. After the heated discussion that can be expected to ensue, the teacher can encourage suggestions for the arithmetical procedures needed to find the correct answer. This can be done in class if the question is used at the start of a period to motivate a review lesson on computation, or it can be assigned as homework if the problem is given at the end of the period to stimulate interest in mathematics.

Students should be led to realize that they have insufficient information because they do not know the thickness of the original piece of paper. At this point tell them to assume that the paper was 0.003 inch thick. Then an estimate of the answer can be obtained as follows.

The correct answer, in inches, is

$$0.003 \times 2^{50} = 0.003 \times 2^{10} \times 2^{10} \times 2^{10} \times 2^{10} \times 2^{10}$$

Since $2^{10} = 1024$, we can estimate this as 1000 and write the product as

$$0.003 \times 1000 \times 1000 \times 1000 \times 1000 \times 1000 = 3,000,000,000,000$$

To change this number of inches to feet and then to miles, we need to divide by 12 and by 5280. As an estimate, divide by 10 and 5000:

$$\frac{3,000,000,000,000}{10 \times 5000} = 60,000,000 \text{ miles}$$

Few students will fail to be astounded at this result! Most will recognize the value of this process as a means of obtaining a reasonable estimate of the correct solution. (To the nearest mile, the actual answer is 53,309,562 miles.)

An interesting related question that can be raised is to ask the class to estimate the number of times that one can physically fold a piece of paper. Most students are extremely surprised to find that it is almost impossible to complete more than seven folding operations. Of course, they assume that one might be able to fare better with a larger piece of paper as a start. An interesting classroom experiment can then be performed. Bring a large sheet of newspaper to class, and have a student attempt to fold it as often as possible. The results prove to be the same, with seven folds remaining the maximum number possible.

Guessing and Estimating

The following items are representative of the many questions that are suitable to invite guesses, promote discussion and estimation, and motivate students to perform the computation necessary to establish answers. Most of the suggestions offered here are particularly worthwhile for use with groups of slow learners in general mathematics classes inasmuch as these youngsters very much enjoy making guesses and can be stimulated to find answers by appropriate computational procedures. In each case, guess first and then compute. (Some of these problems will appear again in the Exercises at the end of this chapter.)

1. What is a set of dimensions for a rectangular box just large enough to hold 1 million pennies?

2. One million people are lined up, with just an arm's length between two adjacent people. How far will this line reach? Will it reach across Pennsylvania? Across the United States from coast to coast? Around the world?

3. A person starts a chain letter. The letter is sent to two people, and each of them is asked in turn to send copies of it to two other people. These recipients are each asked to send copies to two additional people.

Assuming no duplications, how many people in all will have received copies of this letter after the twentieth mailing?

4. How much is a stack of pennies 10 miles high worth?

A number of geometric guessing activities can lead to interesting discoveries. Most of these can be easily verified by the student through some simple construction or experimentation. Following are several representative items in this category; others are considered in Chapter 6.

5. Fold a square piece of paper twice, as shown, and cut off the folded corner. When opened, the paper will have one hole.

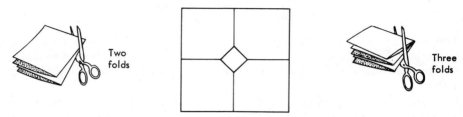

Suppose that the paper is folded three times and the folded corner is cut off. How many holes will there be in the paper? How many holes will the paper have if it is folded four times and then the folded corner is cut off? Does the manner in which the paper is folded affect the answer?

6. Take a piece of paper and fold it down the middle as shown. If you then proceed to tear the paper as indicated, you will get three pieces. Suppose that the paper is folded in half a second time in the other direction before it is torn. How many pieces of paper will there be then?

How many pieces will there be if the paper is folded three times before it is torn? Investigate the various solutions to this problem if the three folds are made in a different manner.

Interesting guessing activities can be provided in the format of multiple-choice questions. As before, allow students to make judicious guesses first and verify answers thereafter.

7. What is the length in inches of a diameter of a penny?

(a) $\frac{3}{8}$ (b) $\frac{1}{2}$ (c) $\frac{5}{8}$ (d) $\frac{3}{4}$ (e) $\frac{7}{8}$

8. Approximately how many pennies are there in 1 pound of pennies?

(a) 100 (b) 150 (c) 200 (d) 250 (e) 300

9. How many pennies must be stacked one on top of the other so that the height of the pile will be equal to the height of a quarter standing on edge?

10. Which of the following segments has a length of 1 inch?

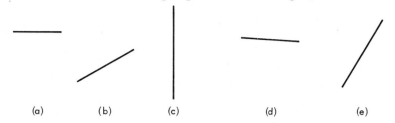

(a)   (b)   (c)   (d)   (e)

11. Which of the following shows the size of a nickel?

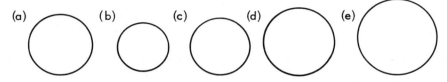

(a)   (b)   (c)   (d)   (e)

2.2   MAKE USE OF "MATHEMAGICAL" NOVELTIES

Many motivational ideas are based on "tricks" that can be justified through relatively simple mathematical procedures. Probably the most common of all of these is the "think-of-a-number" trick. After several operations the teacher is able to determine the number that a person ends up with, but *not* the number originally thought of. Consider, for example, the following simple trick together with its mathematical justification:

THE TEACHER SAYS:	THE TEACHER THINKS:
Think of a number.	n
Double the number.	$2n$
Add 7.	$2n + 7$
Subtract 1.	$2n + 6$
Divide by 2.	$n + 3$
Subtract the number you originally started with.	3

At this point everyone in the class is thinking of the number 3. The teacher can then announce that this number is 3, or can continue with several additional operations to arrive at a special number, such as that day's date. This type of activity is very worthwhile to generate interest in mathematics, but it can specifically be used to introduce a unit of work on variables as well as to show an application of the use of variables in algebra.

The following collection of mathematical tricks is illustrative of the type suitable for use in the classroom for purposes of motivation. In each case, the reader is invited to discover why the trick works as it does. Most of these explanations are within the grasp of average secondary school students, but even where they are not, it is still worthwhile to use the trick as a means of creating and maintaining interest.

Number Tricks

1. INSTRUCTIONS

	EXAMPLE
Write a two-digit number between 50 and 100.	78
Add 54.	+ 54
	132
Cross out the hundreds digit, and add it to the remaining two-digit number.	∤32; 32 + 1 = 33
Subtract your result from the number with which you started.	78 − 33
The result will always be 45.	45

The final outcome depends upon the number added in the second step. Merely subtract this number from 99 to determine the outcome. In this example, 54 was added in the second step, so the result must be 45. If 84 were to be added in the second step instead of 54, the final result would be 15, regardless of the number between 50 and 100 that was chosen at the start.

In each of the following tricks, the participant must give the teacher certain information at a given time. Thereafter the teacher uses this information to determine some special number. Often such tricks are used to guess a person's age, as in the next item.

2. INSTRUCTIONS

	EXAMPLE
Think of your age (or any other number greater than 9).	14
Multiply by 10.	14 × 10 = 140
From this number subtract the product of any one-digit number and 9. That is, subtract some multiple of 9 from 9 through 81.	140 − 27
	113

At this point the teacher asks the student to state the final outcome, 113. To determine the original number, cross out the units digit and add it to the remaining two-digit number. In the example shown, cross out the 3 and then add, to obtain $11 + 3 = 14$, the number started with.

3. INSTRUCTIONS

	EXAMPLE
Think of a number.	21
Multiply by 2.	21 × 2 = 42
Add 5.	42 + 5 = 47
Multiply by 5.	47 × 5 = 235

At this point the student is asked to state the final outcome. To find the original number, merely remove the units digit and subtract 2 from the remaining two-digit number. In the example shown, cross out the 5 and then subtract 2 to obtain $23 - 2 = 21$.

4. INSTRUCTIONS   EXAMPLE

INSTRUCTIONS	EXAMPLE
Think of your age.	15
Multiply your age by 2.	$15 \times 2 = 30$
Add 10.	$30 + 10 = 40$
Multiply by 5.	$40 \times 5 = 200$
Add the number of people in your family.	$200 + 4 = 204$
Subtract 50.	$204 - 50 = 154$

Now ask for the final result. The person's age will be represented by the hundreds and tens digit, the number of people in the family by the units digit. For this example, 154 is interpreted to mean an age of 15 and a family size of 4. (It is assumed the size of the family is not greater than 9.)

5. INSTRUCTIONS   EXAMPLE

INSTRUCTIONS	EXAMPLE
Think of a number.	23
Multiply by 5.	$23 \times 5 = 115$
Add 6.	$115 + 6 = 121$
Multiply by 4.	$121 \times 4 = 484$
Add 9.	$484 + 9 = 493$
Multiply by 5.	$493 \times 5 = 2465$

Ask for the final result, and subtract 165 from this number. Drop the zeros in the units and tens places, and the resulting number is the one started with. In the example shown we have $2465 - 165 = 2300$, and the original number thought of was 23.

Using Aids

Many tricks can appear to be quite dramatic through the use of extra showmanship. For example, the next trick begins with a book of 20 matches and requires some burning.

1. Hand a student a book of 20 matches, and with your back turned give the following directions.

DIRECTIONS	EXAMPLE
Pick any number from 1 through 10 and remove that many matches from the book.	Assume that the student removes 8 matches.
Count the remaining matches and find the sum of the two digits in that number.	12 matches remain. The sum of the two digits is 3.
Remove that many more matches.	The student removes 3 more matches.
Burn some of the remaining matches, one at a time and give them to me.	Assume the student chooses to burn 4 of the remaining matches.
You immediately tell the student how many matches are left in the book.	Subtract 4 from 9 to get 5, the number of matches left in the book.

Regardless of how many matches are removed by the student in the first step, there will be exactly 9 left at the end of the third step. So the teacher needs only to subtract the number of burnt matches returned from 9 to find how many remain in the book.

This trick will work as long as the student begins with a collection of 20 items. Thus if there is an objection to the use of matches in the classroom, substitute 20 toothpicks, chips, or pieces of candy instead. If the student starts with 20 candies and returns those that are left at the end, the teacher can "guess" how many were eaten.

The following trick is an interesting one in that it involves two students.

2. Place a pencil and an eraser on the desk, and ask for two volunteers. (We shall call them John and Mary.)

INSTRUCTIONS	EXAMPLE
One of you take the pencil, and the other take the eraser. Don't tell me which one takes which object; I'll guess it!	Assume that John takes the pencil and Mary takes the eraser.
The one who took the pencil is assigned the number 7; the one who took the eraser has the number 9. (The teacher does not know who has which number at this stage.)	John begins with 7, and Mary begins with 9.
John, multiply your number by 2. Mary, multiply your number by 3.	John: $7 \times 2 = 14$. Mary: $9 \times 3 = 27$.
Find the sum of these products.	$14 + 27 = 41$.

The teacher now asks for this sum. If the sum is divisible by 3, then Mary took the pencil and John took the eraser. If the sum is not divisible by 3, then John took the pencil and Mary took the eraser. In the example shown, the sum of 41 is not divisible by 3, which indicates that John took the pencil.

Card Tricks

Many card tricks that do not involve sleight of hand are based on some mathematical principle. If used judiciously, one of these can be the dramatic highlight of a lesson. The reader is urged to practice with a deck of playing cards before performing the following trick for a class.

Call a student to the front of the room and begin to count out cards in a row on the desk, facedown. The student is instructed to stop your counting at any number between 1 and 9.

Assume you are told to stop after spreading out seven cards; then the number 7 is the student's "magic number."

Next proceed to spread out 20 additional cards on the desk.

Now instruct the student to count backwards to the seventh card, since 7 was the magic number selected. The student picks up this card and shows it to the class.

Much to everyone's surprise you now announce that the card is the king of clubs. (Alternatively, you can place your prediction in a sealed envelope that a student opens and reads to the class.)

This is a trick that works automatically provided you know in advance the twenty-first card from the top of the deck. Thus if you wish the final result to be the king of clubs, merely stack the deck in advance so that this card is in the required position. Can you see why this trick always works?

Note that there is nothing special about the number 20 in the procedure described above. Thus, for example, you can spread out 15 additional cards provided you know the sixteenth card in advance.

2.3 INTRODUCE UNUSUAL ARITHMETIC EXPLORATIONS

It has been said that mathematics is not a spectator sport. That is, it is essential that you get your students involved in the lesson at hand as active participants rather than as passive observers. A rich supply of unusual topics in arithmetic exists that can be used for this purpose. Following is a set of representative items that can be used to motivate students in the mathematics classroom, while providing a review of arithmetic fundamentals and other skills as well.

Finger Computation

Finger computation can be counted on to develop interest in most classes. Multiplication by 9 on one's fingers is intriguing and can be accomplished as shown in the following diagrams.

To multiply 3×9, bend down the third finger from the left. Then read the answer in groups of fingers on either side of the bent finger.

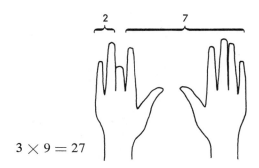

$3 \times 9 = 27$

Here are several additional examples:

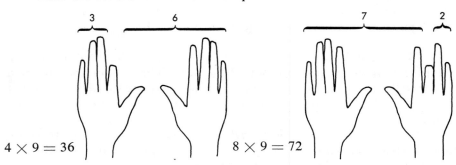

$4 \times 9 = 36$ $8 \times 9 = 72$

Fingers can be used to multiply a two-digit number by 9 provided that the units digit is greater than the tens digit. To multiply 9×38, first put a space after the third finger from the left. Then bend the eighth finger from the left. Read the answer in terms of groups of fingers.

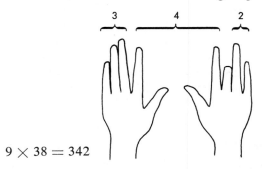

$9 \times 38 = 342$

Magic Numbers

1. Not quite as spectacular, but nevertheless interesting, is to walk into class and place on the board the "magic number" for the day:

$$12,345,679$$

When the class asks what is magic about this number, the teacher proceeds to ask row 1 to multiply the number by 9. Row 2 is to multiply the same number by 18 (**2** × 9), row 3 by 27 (**3** × 9), row 4 by 36 (**4** × 9), and row 5 by 45 (**5** × 9). The following interesting products are obtained:

$$9 \times 12,345,679 = 111,111,111$$
$$18 \times 12,345,679 = 222,222,222$$
$$27 \times 12,345,679 = 333,333,333$$
$$36 \times 12,345,679 = 444,444,444$$
$$45 \times 12,345,679 = 555,555,555$$

This same pattern continues for the remaining multiples of 9 through 81.

Obviously there are many ways in which any of these examples may be used. For example, as an alternative approach the teacher asks the class if anyone has a favorite or lucky number between 1 and 10. The student who responds with 3 as a lucky number is asked to multiply 12,345,679 by 27 and report on the result; the student who offers 7 as the lucky number is told to multiply 12,345,679 by 63; and so forth. The surprise on the students' faces as they complete the multiplication is rewarding and can be contagious for the rest of the class as well!

2. "Magic numbers" can be used to fascinate students and motivate them to review arithmetic in a disguised form. As another example, the magic number 15,873 works with multiples of 7. If 15,873 is multiplied by $7 \times n$, where n is a number from 1 through 9, the product consists of repeated n's. An interesting way to use this item in a class is to ask a student to select a favorite number from 1 through 9. Assume the student responds with 8. Then ask the class to multiply 15,873 by 56 ($7 \times$ **8**); the product will be 888,888. Here are several additional examples:

$$\begin{array}{ccc}
\begin{array}{r}15{,}873 \\ \times\ 14 \\ \hline 222{,}222\end{array} \quad (7 \times \textbf{2})
&
\begin{array}{r}15{,}873 \\ \times\ 28 \\ \hline 444{,}444\end{array} \quad (7 \times \textbf{4})
&
\begin{array}{r}15{,}873 \\ \times\ 49 \\ \hline 777{,}777\end{array} \quad (7 \times \textbf{7})
\end{array}$$

There are other such numbers that can be used in a similar fashion. For example, the number 8547 works with multiples of 13 as shown here:

$$\begin{array}{ccc}
\begin{array}{r}8547 \\ \times\ 26 \\ \hline 222{,}222\end{array} \quad (13 \times \textbf{2})
&
\begin{array}{r}8547 \\ \times\ 39 \\ \hline 333{,}333\end{array} \quad (13 \times \textbf{3})
&
\begin{array}{r}8547 \\ \times\ 65 \\ \hline 555{,}555\end{array} \quad (13 \times \textbf{5})
\end{array}$$

Secret Messages

Students often enjoy sending a secret message based on a set of computational procedures. The following directions will produce the message "We like math."

DIRECTIONS	EXAMPLE
Write any three-digit number so that the hundreds digit is at least two more than the units digit.	582
Reverse the digits and subtract.	$\begin{array}{r}\ \ \ 582\\ -\ 285\\ \hline 297\end{array}$
Reverse the digits and add. (This sum will always be 1089.)	$\begin{array}{r}297\\ +\ 792\\ \hline 1089\end{array}$
Multiply by 10,000,000 and subtract 1,257,138,525.	$\begin{array}{r}10{,}890{,}000{,}000\\ -\ 1{,}257{,}138{,}525\\ \hline 9{,}632{,}861{,}475\end{array}$

Convert this difference into letters, using this key:

For each 1, write M.
For each 2, write I.
For each 3, write L.
For each 4, write A.
For each 5, write H.
For each 6, write E.
For each 7, write T.
For each 8, write K.
For each 9, write W.

9	6	3	2	8	6	1	4	7	5
W	E	L	I	K	E	M	A	T	H

(See page 158 for an algebraic explanation as to why the intermediate result will always be 1089.)

Using a Calculator

To get all the students into the act, have each one enter a three-digit number, such as 735, on a calculator. Then repeat the same three digits to form a six-digit number, such as 735,735. The students are then instructed to divide this number by 7, by 11, and then by 13 in succession. They should register surprise when they find that the final result is their original three-digit number.

After demonstrating this "trick," the teacher should encourage the students to search for some explanation as to why it works. As a hint, if necessary, suggest that students find the product $7 \times 11 \times 13$. This product is equal to 1001, and $(1000 + 1)$ times any three-digit number will give a six-digit number with the first three digits repeated.

Shortcuts

Most students find shortcuts to computation quite interesting, and many of these exist. One particularly interesting shortcut enables students to quickly square a number whose units digit is 5. For example:

$$35^2 = 1225 \quad (3 \times 4 = 12)$$
$$45^2 = 2025 \quad (4 \times 5 = 20)$$
$$55^2 = 3025 \quad (5 \times 6 = 30)$$

Note that each product ends with the last two digits as 25. The first two digits represent the product of the tens digit of the original number and its successor. That is, consider the square of 75:

$$75^2 = \underbrace{56}_{7 \times 8}25$$

Can you show why this trick works? Try to find additional examples of shortcuts suitable for use in the classroom.

2.4 MAKE USE OF GEOMETRIC CHALLENGES

Every teacher should have his or her own special collection of geometric tidbits—short little puzzles, problems, and curiosities in geometry to warm up the class, to gain attention, to involve, to challenge, to maintain interest, or simply to give a change of pace.

Consider, for example, the following item that can also serve as the problem of the day or week. Two line segments are drawn with their endpoints as points on the triangle. Then count the number of sides of the resulting regions and find their sum, as in these figures:

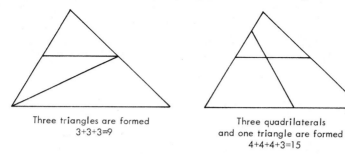

Three triangles are formed
3+3+3=9

Three quadrilaterals
and one triangle are formed
4+4+4+3=15

Now the challenge is to draw other figures to represent each of the numbers from 10 through 14. It is an interesting challenge because, for no special reason, it turns out to be impossible to provide a representation for 12! Occasionally it is useful to have students attempt impossible problems so as to develop the notion that not everything can be established or proved.

The geometric examples that follow are but a few of the vast supply available in the literature or ready for the teacher to create. They are ideas for motivation as compared to more extensive classroom activities and experiments such as those found in later chapters. The particular use of these ideas in the mathematics class is left to the imagination of the teacher.

1. Move just three dots to form an arrow pointing down instead of up.

2. Form four equilateral triangles with just six toothpicks.

3. How many pennies can you arrange such that each penny touches every other penny?

4. To mount a picture, two thumbtacks are needed in any two corners. What is the least number of tacks needed to mount four pictures?

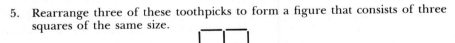

5. Rearrange three of these toothpicks to form a figure that consists of three squares of the same size.

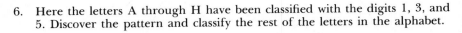

6. Here the letters A through H have been classified with the digits 1, 3, and 5. Discover the pattern and classify the rest of the letters in the alphabet.

1	A E F H
3	C
5	B D G

7. Without lifting your pencil from the paper, try to draw four connected lines that pass through all nine points. Remember, the lines must be straight.

8. How many squares are in this figure?

9. How many rectangles are in this figure?

2.5 USE BULLETIN BOARD DISPLAYS TO GENERATE INTEREST

In many classrooms the bulletin board serves as the focal point for exhibits and posters. In others it serves, as does the blackboard, in the actual teaching and on-the-spot development of an idea. In some it displays daily mathematical tidbits, historical quotes, or puzzle problems. In still others it is the source of enrichment materials related to the unit on hand. Conveniently available in most classrooms, the bulletin board can serve an important role in the overall learning experience. To illustrate the potential of the bulletin board, several illustrations of possible uses are given here. In general, its use is restricted only by the imagination of the teacher using it.

Optical Illusions

A good topic for a bulletin board display in the mathematics classroom that almost everyone enjoys is optical illusions. They can be effectively exhibited just for fun as well as to convince students that seeing is not always believing. This can be an important point to make in geometry where students are encouraged in their proofs to rely on reason rather than relying too heavily on figures.

Some of the many popular illusions are given in this collection. A new one might be added each week and students encouraged to search for more on their own. These should be placed on the bulletin board to stimulate student discussion and reactions. Ask students to describe what they see.

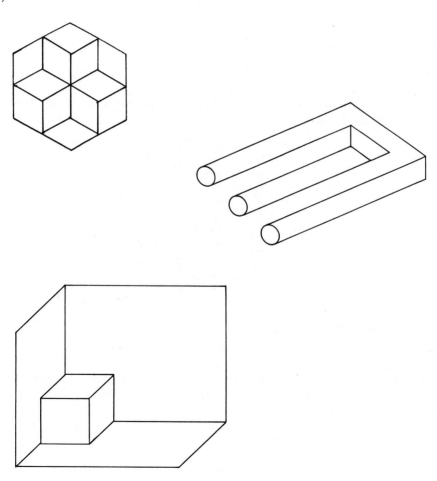

For the following, ask students to compare the lengths of segments (a) and (b) visually. Then allow them to measure the segments to confirm their impressions.

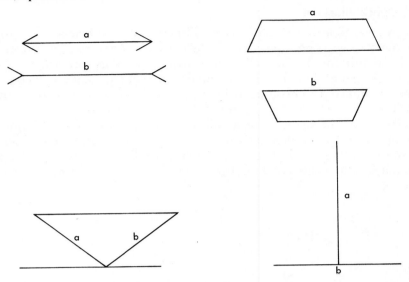

Of course, when the optical illusions are taken down, they should be stored in the box shown here:

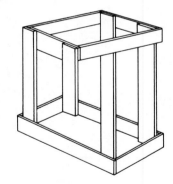

Student Projects

Students, or groups of students, should be encouraged to prepare bulletin board displays themselves. Individual posters made by the teacher or student can bring life to the bulletin board as well. They should be eye-catching, with vivid colors, snappy titles, and simply stated questions, such as the following:

REFLECTIONS

READ THIS MESSAGE SEEN IN
A MIRROR. CAN YOU FIND THE
EIGHT MISTAKES IN IT?

DART TIME

FIND ALL THE SCORES POSSIBLE
WITH JUST **4** DARTS

Here is a special project on tessellations made by a student and effectively displayed on the bulletin board.

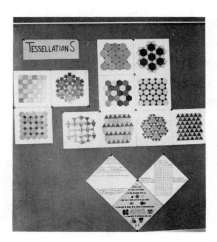

Historical Items

In a slightly different vein, interesting bits from the history of mathematics can be presented via the bulletin board. Stories about Thales, Pythagoras, Euclid, and Archimedes make interesting reading when studying geometry. Or pictures of Descartes, Pascal, and Gauss can be posted when studying coordinate geometry, probability, and complex numbers. Have students interested in philately search out and display some of the stamps relating to mathematics and mathematicians. It is surprising how exciting this can be to some students.

In the illustrations of stamps shown here, the Scott catalog numbers are given in parentheses for each stamp. All the stamps shown are reasonably inexpensive.

Reise (Germany, 779)

Ramanujan (India, 369)

Stevin (Belgium, B321)

Einstein (USA 1285)

Pascal (France, B181)

Bolyai (Romania, 1345)

Chebyshev (Russia, 1050)

Descartes (France, 330)

Ask your students to identify the figure or formula represented on each of these stamps. Reproduce one or more of these stamps of mathematicians and see if your students can identify the major contributions of each.

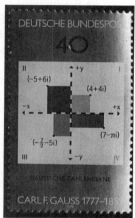

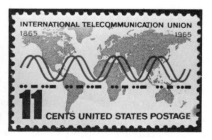

As a Teaching Aid

The bulletin board can also function strictly as an immediate aid in teaching as illustrated with this example on regular tetrahedrons. Congruent equilateral triangles measuring about 2 inches on each edge are prepared. Students then set up the various possible patterns or nets of four triangles by tacking triangles on the bulletin board. Only three possible arrangements exist if rotations and reflections are not counted. And only two of those form patterns for a tetrahedron. Which two?

Exploration

Interesting discovery problems can also be illustrated on the bulletin board. One challenging example is the shortest-path problem. Three tacks are placed on the bulletin board in a triangular array and a reasonable distance apart. Equipped with a string and a ruler, students try to find the shortest path connecting the points.

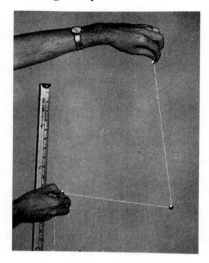

The problem is not as simple as it may sound, but this will make the students only more anxious to try. The solution is somewhat surprising. If the triangle itself contains an angle of 120° or more, the two shortest sides of the triangle form the best path. Otherwise, an intermediate point that forms three 120° angles with the given points locates the shortest connecting path.

As an extension to this problem, consider the shortest path connecting the four vertices of a square.

2.6 DISCUSS APPLICATIONS OF MATHEMATICAL CONCEPTS

One of the most embarrassing questions faced by any teacher of mathematics is, "What good is all of this?" Unfortunately, for some topics, we may have to resort to standard answers, such as indicating the future value of the topic in the student's study of mathematics. However, it enhances motivation if we provide meaningful applications of topics under study. Consider, for example, a unit on the conic sections where the ellipse is being studied.

You can tell the story of the whispering gallery in Washington, D.C. When a person stands at one of the foci of the ellipse and whispers, regardless of the noisy condition of the chamber, someone standing at the other focus can hear the whispered message quite clearly.

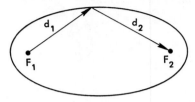

The explanation of this phenomenon is that all sound waves coming from one of the foci will bounce off the sides of the ellipse and pass through the other focus. By the definition of an ellipse, the sum of the two distances in the figure, $d_1 + d_2$, is constant.

At one time one of the authors indicated to an algebra class that this property would apply to an elliptical pool table. Thus if a ball is placed at F_1 and hit hard *in any direction* it is guaranteed that the ball will hit another one placed at F_2. After class one student indicated that this was the most interesting thing that he had ever learned in all his studies of mathematics!

Before leaving this topic, it should be noted that what is considered to be a genuine application by the teacher is not necessarily one for the student. Thus a student may be quite satisfied to see how mathematics is used to justify a trick or a shortcut, and will be content that this is an application of a topic under study. For example, consider an algebra class that has studied the formula

$$(a - b)(a + b) = a^2 - b^2$$

A real application of this formula for many students is its use to multiply a product such as 78×82 mentally:

$$78 \times 82 = (80 - 2)(80 + 2) = 80^2 - 2^2 = 6396$$

Likewise, students will accept the mathematical explanation of a mathematical trick, such as the one shown on page 31, as a suitable application of a mathematical topic.

Finding suitable applications of mathematics is not a trivial task and is easier for some topics than for others. Here are just a few examples of interesting applications to discuss in class:

1. In the event of an accident, police may estimate the speed of an automobile by measuring the length of the skid marks left when the driver was braking. Thus the speed in miles per hour, s, can be approximated by the formula $s = \sqrt{24d}$, where d is the length of the skid marks in feet. (The constant, 24, will vary in adverse road and weather conditions.)

2. It is known that the air temperature decreases about 5°F for every 1000 feet of altitude. Meteorologists use the following formula to show this relationship:

$$T = t - 5\left(\frac{A}{1000}\right)$$

In this formula, t represents the ground temperature, A is the altitude in feet, and T is the temperature at altitude A, in degrees Fahrenheit.

3. A parabolic surface can be formed by rotating a parabola about its axis. When a source of light is placed at the focus, all light from this source will be reflected in parallel rays, as evidenced by the parabolic reflectors used in headlights and searchlights.

4. It has been determined that the rectangle most pleasing to the eye is one whose dimensions are such that the width is approximately 0.618 times the length. This ratio is referred to as the *golden ratio,* and the rectangle as the *golden rectangle.* Using w as the width and l as the length, the golden rectangle satisfies the following proportion:

$$\frac{w + l}{l} = \frac{l}{w}$$

If $l = 1$, this proportion can be solved to show that $w = 0.618$. The dimensions of many famous works of art and historic buildings (such as the Parthenon) are based on this ratio. (See page 186 for a further discussion of the golden ratio.)

2.7 CONCLUSION

In a sense there is no real conclusion to the topic of motivation. The items cited in this chapter represent only a small subset of the total ways and means of motivating students. All the material in the remainder of this text may be considered as dealing in some way with the motivation of students in mathematics.

Above and beyond all else, suitable motivation depends upon the enthusiasm and imagination of the teacher. When all is said and done, it is the teacher who counts, and no motivational technique is of value if the teacher does not display his or her genuine interest in the subject at hand.

EXERCISES

1. What are the dimensions of a rectangular box that would be large enough to hold 1 million pennies?
2. A person starts a chain letter. The letter is sent to two people, and each of them is asked in turn to send copies of it to two other people. These recipients

are each asked to send copies to two additional people. Assuming no duplications, how many people in all will have received copies of the letter after the twentieth mailing?

3. What is the diameter of a penny in inches?

   (a) $\dfrac{3}{8}$ (b) $\dfrac{1}{2}$ (c) $\dfrac{5}{8}$ (d) $\dfrac{3}{4}$ (e) $\dfrac{7}{8}$

4. Approximately how many pennies are there in 1 pound of pennies?

   (a) 100 (b) 150 (c) 200 (d) 250 (e) 300

5. The length of a $1 bill is about how many times its width?

   (a) $\dfrac{5}{3}$ (b) $\dfrac{6}{3}$ (c) $\dfrac{7}{3}$ (d) $\dfrac{8}{3}$ (e) $\dfrac{9}{3}$

6. About how many pennies would have to be piled one on top of another to reach the ceiling of a room that is 8 feet high?

7. How much is a stack of pennies 10 miles high worth?

8. To the nearest thousand, how many pennies are there in 1 mile of pennies placed next to each other with their edges touching?

9. How long would it take to spend $1 million at the rate of $100 every minute?

10. Provide a mathematical explanation as to why the match trick illustrated on page 33 works.

11. Explain why the card trick discussed on page 34 works.

12. Draw within a triangle any two line segments such that their endpoints are points on the triangle. Count the number of sides of the resulting regions and find their sum. Draw figures that illustrate the sums of 10, 11, 13, and 14. (See page 39.)

13. Try the following trick on a friend; then provide a mathematical justification to show why it works. Have your friend roll a die three times and record the results in order. Then give these instructions:

    Multiply the first number by 2.
    Add 5 to the result, and then multiply by 5.
    Add the second number to this product.
    Multiply by 10 and add the third number.

    You then ask your friend to state the final result. From the number mentally subtract 250. The result will be a three-digit number whose digits represent the original three numbers. (Instead of tossing three dice, you can merely have your friend think of a three-digit number.)

14. Find the shortest path connecting the four vertices of a square.

ACTIVITIES

1. Prepare a set of multiple-choice questions similar to those found on page 30 that requires one to guess and estimate about familiar objects.

2. Prepare a collection of magic tricks that are based on some mathematical principle. Demonstrate at least one of these in class.

3. Search for other examples of the use of fingers in completing computations.

4. Prepare a collection of various computational shortcuts. Demonstrate at least one of these in class.

5. Prepare a bulletin board display suitable for a junior high school mathematics class.

6. Repeat Activity 5 for a senior high school class.

7. Prepare a collection of at least 10 applications of mathematics suitable for a junior high school class.

8. Repeat Activity 7 for a senior high school class.

9. Examine at least two junior high school textbooks and report on the use made of motivational techniques in each. In particular, report on any unusual approaches to motivation that are used.

10. Repeat Activity 9 for two senior high school mathematics books.

11. Repeat Activity 9 for a textbook that is specifically designed for slow learners in mathematics.

12. Develop a one-period lesson plan for a junior high school mathematics class that clearly indicates procedures to be used to motivate the lesson.

13. Repeat Activity 12 for a senior high school mathematics class.

14. Prepare a collection of at least 10 geometric challenges similar to those given in section 2.4.

READINGS AND REFERENCES

1. Read about mathematics used on postage stamps in *Mathematics and Science, An Adventure in Postage Stamps* by William L. Schaaf, a publication of the National Council of Teachers of Mathematics (1978).

2. Read about applications of mathematics in the 1980 publication of the National Council of Teachers of Mathematics titled *A Sourcebook of Applications of School Mathematics*. In particular, report on the chapter by Pamela Ames titled "A Classroom Teacher Looks At Applications," as well as on several of the specific applications described from the areas of algebra and geometry.

3. The 1986 Yearbook of the National Council of Teachers of Mathematics is titled *Estimation and Mental Computation*. Read and report on Chapter 17, " 'Guess and Tell': An Estimation Game" by Beth M. Schlesinger.

4. Obtain a book that deals with magic tricks with cards and report on several of these that are based on mathematical principles. Among those to read is *Magic With Cards* by Frank Garcia and George Schindler, New York: David McKay Company, Inc., 1975. Also see *Self-Working Card Tricks* by Karl Fulves, New York: Dover Publications, Inc., 1976.

5. Prepare a report on finger computation. Among the sources to examine is the 31st Yearbook of the National Council of Teachers of Mathematics titled *Historical Topics for the Mathematics Classroom*, 1969. In particular, read Capsule 29, "Finger Reckoning." Also see Chapter 8, "Finger Arithmetic" in *Mathematical Magic Show* by Martin Gardner, New York: Alfred A. Knopf, 1977.

6. Read and report on three of the chapters in the 1979 Yearbook of the National Council of Teachers of Mathematics, titled *Applications in School Mathematics*.

7. Read the 1978 publication of the National Council of Teachers of Mathematics titled *Mathematics and Humor* by Aggie Vinik, Linda Silvey, and Barnabas Hughes.

8. Read Chapter 13, "Large Numbers and the Calculator" by William B. Fisher

and Jim N. Jones in the 1982 Yearbook of the National Council of Teachers of Mathematics titled *Mathematics for the Middle Grades* (5–9).

9. The 1977 Yearbook of the National Council of Teachers of Mathematics is titled *Organizing for Mathematics Instruction*. Read and report on Chapter 13, "Hand-held Calculators: Past, Present, and Future" by Max Bell, Edward Esty, Joseph N. Payne, and Marilyn N. Suydam.

10. Read and report on three of the calculator activities presented in *Calculators*, a 1979 publication of the National Council of Teachers of Mathematics that presents readings from their journals.

Motivating Problem-Solving Instruction

Chapter 3

At the start of the decade of the 1980s the National Council of Teachers of Mathematics published a document entitled *An Agenda for Action: Recommendations for School Mathematics of the 1980s*. Designed to serve as a blueprint for change, this document has served as a guide to both textbook authors and numerous communities and states throughout the country for revising the mathematics curriculum. Its first recommendation is the one that has received the widest attention and acclaim:

Problem solving must be the focus of school mathematics in the 1980s.

As a result of this NCTM recommendation, as well as widespread interest in problem solving by mathematics educators, the subject became the major topic of discussion throughout the decade at various professional meetings. There is no doubt that we must continue to focus attention on strategies for problem solving since our students give evidence on local and national tests that they have much more difficulty in this area than they do in basic computational skills. However, we must also note that students are not nearly as excited about problem solving as are their teachers. Thus it is essential that we search for ways to motivate problem-solving instruction; we need to develop student skills in solving problems, but first we need to generate interest in this topic.

There are many lists of problem-solving strategies available. In this chapter we select several of these strategies, and offer suggestions for each that can help motivate student interest. Often these strategies are best illustrated through the use of nonroutine problems.

3.1 STRATEGY FOR PROBLEM SOLVING: TRIAL AND ERROR

Some problems are best solved by trial and error, with logical thinking accompanying the process. One suitable way to initiate students in this approach is to offer a famous historical problem, such as the following.

The Koenigsberg Bridge Problem

It is claimed that the inhabitants of the ancient town of Koenigsberg enjoyed taking walks over the seven bridges that connected the mainland to two islands in the river that ran through the town.

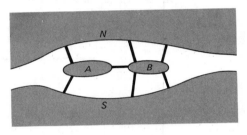

The problem that they considered was whether they could start at any point and take a walk that would cross each bridge exactly once, without retracing their steps. This is an interesting problem to give to junior high school students as a home assignment "just for fun." They will undoubtedly be extremely frustrated when you tell them the following day that such a walk is impossible! On the other hand, part of the strategy of trial and error is to try to find a solution by exhausting all possibilities. The solution to this particular problem is that there is no such path that crosses each bridge once and only once.

Traversable Networks

As a follow-up to the Koenigsberg Bridge problem, suggest that your students prepare a report on the work of the Swiss mathematician Leonhard Euler concerning networks. A network on a plane consists of points called *vertices*, paths connecting them called *arcs*, and *regions* bounded by them. Euler developed methods for determining whether such networks can or cannot be traced with a continuous path. For example, the two networks shown on the next page at the left can be traversed without lifting your pencil from the paper or retracing any path more than once, whereas the two at the right are not traversable under these conditions.

Traversable   Not traversable

In the preceding figures, an *odd vertex* is one that has an odd number of arcs at that point; an *even vertex* has an even number of arcs. Euler proved that such networks can be traced if one of these two conditions is met:

1. There are only even vertices in the figure.
2. There are exactly two odd vertices in the figure.

In the latter case, one must start at one of these odd vertices and the path will terminate at the other odd vertex. Encourage students to test Euler's conditions with figures of their own, and to draw a network that illustrates the Koenigsberg Bridge problem.

Regardless of whether a network can be traversed or not, Euler discovered an interesting relationship concerning such figures. He showed that the number of vertices plus the number of regions is always two more than the number of arcs.

$$V + R = A + 2$$

(Euler also developed a corresponding formula for polyhedrons; see page 204.)

As a classroom experiment, provide your students with a set of figures such as the following:

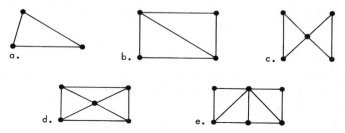

Have the students count the number of vertices, regions, and arcs for each of the networks shown and record the results in a table.

NETWORK	VERTICES	NUMBER OF REGIONS	ARCS
(a) (b) (c) (d) (e)			

Now ask questions such as these:

1. For each of these networks is the number of arcs *A* ever less than the number of vertices *V* or the number of regions *R*? Is *A* less than the sum of *V* and *R*? Can you discover a relationship among the number of arcs *A* and the number of vertices and regions *V* and *R*? Try to express this relationship in symbols.

2. Now count *V*, *R*, and *A* for each of these networks. Do they support your formula from step 1?

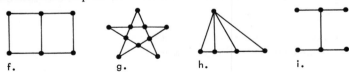

f. g. h. i.

Magic Triangles

Let us consider one more example of a problem that can be approached by trial and error, or by combining that method with logical reasoning. The problem is to place the numerals 1 through 9 in a triangular array so that the sum of the numbers along each side of the triangle is the same. The solutions may not be unique, but here are illustrations of arrays that produce sums of 17 and 23 along each of the sides:

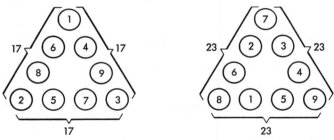

The problem now is to display such triangular arrays for sums between 17 and 23 using the same numbers 1 through 9. Students can be asked to attempt a solution strictly by a trial-and-error approach. An alternative approach is to consider a logical development, noting that the sum of the corner numbers for the first triangle is 6 (1 + 2 + 3), and for the second triangle is 24 (7 + 8 + 9).

SUM OF NUMBERS ALONG THE SIDES	SUM OF CORNER NUMBERS		
17	6		If we consider the difference
18	?	(9)	24 − 6 = 18, and divide this
19	?	(12)	into six equal intervals, we
20	?	(15)	have a reasonable guess for
21	?	(18)	the sums of corner numbers.
22	?	(21)	
23	24		

Using the sums of corner numbers indicated in the table, and a trial-and-error strategy, it proves possible to find triangular arrays for sums of 19, 20, and 21. Unfortunately, no such triangular arrays are possible for sums of 18 and 22.

The very fact that the preceding problem is possible to solve for certain numbers, and not for others, usually motivates students to explore other solutions by using the valuable strategy of trial and error.

3.2 STRATEGY FOR PROBLEM SOLVING: USE AN AID, MODEL, OR SKETCH

Very often a problem can best be solved or at least understood by drawing a sketch, folding a piece of paper, cutting a piece of string, or making use of some other simple and readily available aid. The strategy of using an aid can make a situation real to a student, help to motivate her or him, and generate interest in the problem. Here are several problems that illustrate this strategy.

1. You have five coins: a penny, a nickel, a dime, a quarter, and a half-dollar. You also have two identical boxes. How many different ways can these coins be distributed into the two boxes so that at least one coin is in each box?

This is a suitable problem for elementary or junior high school students to solve by actually drawing a sketch to show all logical possibilities. In the following, the five coins are designated by the numerals 1, 5, 10, 25, and 50. Note that the diagrams indicate a systematic way of enumeration so as to avoid omitting any possibilities.

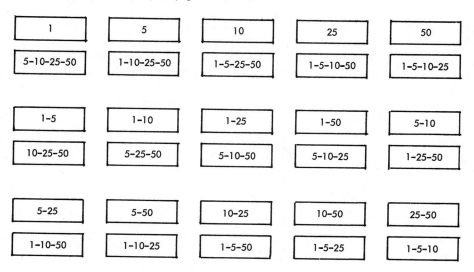

By actually listing all possibilities, we find the solution to be 15.

2. As another example, consider the following problem that is a standard in most elementary calculus textbooks.

An open box, without a top, is to be formed from a 9 × 12-inch piece of cardboard by cutting off square corners of equal size and folding up the edges. What should be the size of the square corners cut off in order to obtain a box of maximum volume?

This problem, solved formally in calculus texts, can be solved informally and experimentally in a junior high school class. Either as a demonstration by the teacher, or experimentally by groups of students in a laboratory setting, use at least four pieces of cardboard of the same size and cut off square corners with dimensions of 1 inch, 2 inches, 3 inches and 4 inches. Use Scotch tape to hold the sides together and allow students to guess which box appears to have the maximum volume before proceeding with any calculations. Many students will guess that all have the same volume.

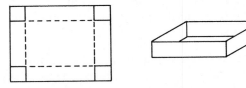

Now, by computation, we can find the volume of each box and summarize the data in a table such as the following:

BOX	SIZE OF SQUARE CUT FROM CORNER	LENGTH-WIDTH-HEIGHT IN INCHES	VOLUME IN CUBIC INCHES
A	1″	10 × 7 × 1	70
B	2″	8 × 5 × 2	80
C	3″	6 × 3 × 3	54
D	4″	4 × 1 × 4	16

Graphing the volume against the size of the corner squares provides further visualization of these results. Connecting the dots on the graph gives a general picture of the relationship involved, but does not indicate exactly what size squares will result in a box of the maximum volume.

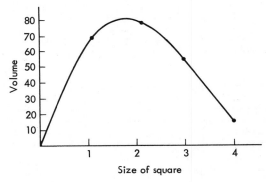

At this point, students should be ready to see that we can test corner squares of $1\frac{1}{2}$ inches and $2\frac{1}{2}$ inches for a better approximation. The respective volumes are then found to be 81 and 70 cubic inches. Thus square corners of $1\frac{1}{2}$ inches prove to be our best estimate for obtaining the maximum volume. (By the methods of calculus, the solution is $\dfrac{7-\sqrt{13}}{2}$, or approximately 1.7 inches.)

3. Drawing a sketch sometimes sheds new light on a problem by exposing alternate approaches. Consider this problem dealing with successive slices in a cake.

One slice through a cake gives two pieces. Two slices can give as many as four pieces. What is the maximum number of pieces you can get with three successive slices? What about four successive slices?

Quick sketches verify the first two cases and offer a possible solution to the third.

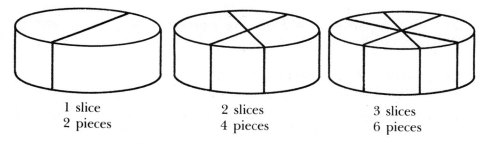

| 1 slice | 2 slices | 3 slices |
| 2 pieces | 4 pieces | 6 pieces |

But there are no restrictions given on the shape of the pieces so the third slice doesn't necessarily have to follow in the same fashion. Can a third slice be repositioned to yield more pieces? Here are two alternate choices that yield 7 and 8 pieces, respectively.

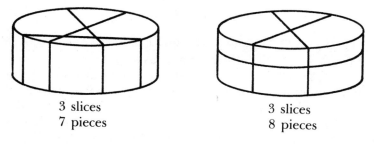

| 3 slices | 3 slices |
| 7 pieces | 8 pieces |

Starting with the sketch on the right, where would you cut with the next slice to get the maximum number of pieces with four slices? Two possibilities immediately come to mind. How many pieces do you get in each of these cases?

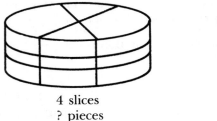 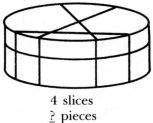

4 slices   4 slices

? pieces   ? pieces

The figure on the left gives 12 pieces while the one on the right gives 14. But there is a better choice for 4 slices—one that gives even more pieces. Can you reposition the fourth slice for the maximum number of pieces possible? (See page 171 for further discussion on the number sequence generated by this problem. But don't turn there until you give this problem some more thought here by sketching other possibilities.)

Drawing a picture or using an aid not only serves as an effective strategy for problem solving in many situations, but also helps to motivate the student to pursue an otherwise vague problem. Many other examples of the use of aids are presented in the following chapters.

3.3 STRATEGY FOR PROBLEM SOLVING: SEARCH FOR A PATTERN

Searching for patterns and then forming generalizations is a very powerful problem-solving strategy that will be explored again in detail in a later chapter (see page 115). Once again, however, we need to search for suitable problems that will create interest on the part of students and thus motivate them to make use of this strategy. One interesting problem that can set the stage for an exploration of this strategy is to challenge your students to find the pattern in the following set of numbers:

8, 5, 4, 9, 1, 7, 6, 3, 2

This is a good problem to give as the Problem of the Week, and then not disclose the solution until sufficient time has elapsed for all to try it. Many students may appear frustrated when they learn the answer: The numbers are arranged in alphabetical order according to their spellings!

A good teaching technique is to have students provide their own nonstandard sequences of numbers for the class to discover. For example, in a city such as New York with numbered subway stops, they may use successive subway stops. Challenge them to use their imagination.

Problems that involve searching for a pattern are plentiful, but selecting those that will stimulate the interest of students is a challenging task. The following problems have proved useful for motivating students to search for patterns.

1. In the following decimal, how many 2's are there in all before the hundredth 3?

$$0.23223222322223\ldots$$

Here the student should note that there is one 2 before the first 3, two 2's before the second 3, and in general n 2's before the nth 3. Thus, before the hundredth 3, the number of 2's is

$$1 + 2 + 3 + 4 + \ldots + 98 + 99 + 100$$

The total number is then found to be 5050, using the method of Gauss. (See page 5.)

2. What is the units digit in the power of 3 with an exponent corresponding to the current year. For example, what is the units digit in the expansion of 3^{1989}? By expanding some powers of 3 we can observe a pattern in successive units digits:

$$
\begin{aligned}
3^1 &= 3 \\
3^2 &= 9 \\
3^3 &= 27 \\
3^4 &= 81 \\
3^5 &= 243 \\
3^6 &= 729 \\
3^7 &= 2187 \\
3^8 &= 6561 \\
\end{aligned}
$$

. . . .

The pattern formed by the units digits is

$$3, 9, 7, 1, 3, 9, 7, 1, \ldots$$

Thus, when n is an integer, 3^{4n} will have a units digit of 1. The number 1988 is a multiple of 4.

$$3^{1988} = 3^{4(497)}$$

So 3^{1988} will have a units digit of 1. The number 3^{1989} is the next successive multiple of 3 and thus, by the pattern observed, will have a units digit of 3. Following this line of reasoning,

3^{1990} will have a units digit of 9,
3^{1991} will have a units digit of 7,
3^{1992} will have a units digit of 1 and so on.

A more difficult and challenging problem is to determine the tens digit in the expansion of 3^{1989}. (See Exercise 4.)

3. In all, how many squares of all sizes are there in a standard checkerboard? Since this problem would be too difficult to solve by actually counting, students should be urged to search for a pattern. Also, they

should be reminded of the valuable strategy of attempting a similar, but less complex, problem. Thus, they might count the number of squares in smaller grids of 1 x 1, 2 x 2, 3 x 3, and 4 x 4.

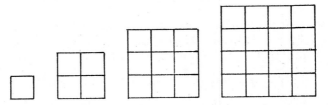

It is helpful to summarize the observed data in a table.

SIZE OF GRID	NUMBER OF SQUARES OF DIFFERENT SIZE				TOTAL
	1 × 1	2 × 2	3 × 3	4 × 4	
1 x 1	1	—	—	—	1
2 x 2	4	1	—	—	5
3 x 3	9	4	1	—	14
4 x 4	16	9	4	1	30

The pattern appears to involve the sums of squares. Thus for a checkerboard of size 8 x 8, the total number of squares is

$$1 + 4 + 9 + 16 + 25 + 36 + 49 + 64 = 204$$

Detailed solutions are not provided for the problems that follow. Each will appear again in the Exercises at the end of this chapter.

4. A number is formed by writing the counting numbers in order, as follows:

$$123456789101112131415161718192021\ldots$$

What is the one millionth digit in this number? (You may wish to first try to find the one hundredth digit and note the process and patterns used.)

5. What is the sum of the first one hundred odd numbers? The first one hundred even counting numbers?

6. Consider the following array of fractions:

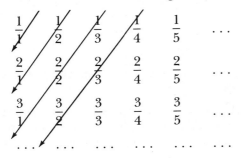

Assume this array extends indefinitely in both directions.

Now arrange the fractions along the indicated diagonals, and match each fraction with a counting number in this manner:

$$\frac{1}{1} \quad \frac{1}{2} \quad \frac{2}{1} \quad \frac{1}{3} \quad \frac{2}{2} \quad \frac{3}{1} \quad \frac{1}{4} \quad \frac{2}{3} \quad \frac{3}{2} \quad \frac{4}{1} \quad \cdots$$

$$\updownarrow \quad \updownarrow \quad \updownarrow \quad \updownarrow \quad \updownarrow \quad \updownarrow \quad \updownarrow \quad \updownarrow \quad \updownarrow \quad \updownarrow$$

$$1 \quad 2 \quad 3 \quad 4 \quad 5 \quad 6 \quad 7 \quad 8 \quad 9 \quad 10$$

If this array is continued, which counting number would be matched with the fraction $\frac{15}{21}$? Which number would be matched with the fraction $\frac{p}{q}$? As a hint, note the pattern of sums of the numerator and denominator of each fraction along each diagonal:

Diagonal	$\left(\frac{1}{1}\right)$	$\left(\frac{1}{2}, \frac{2}{1}\right)$	$\left(\frac{1}{3}, \frac{2}{2}, \frac{3}{1}\right)$	$\left(\frac{1}{4}, \frac{2}{3}, \frac{3}{2}, \frac{4}{1}\right)$	$\cdots$
Sum	2	3	4	5	$\cdots$

This particular problem is appropriate for a senior high school advanced mathematics class. It is related to the German mathematician Cantor's attempt to prove that the set of all rational numbers can be matched in a one-to-one correspondence with the set of counting numbers.

3.4 STRATEGY FOR PROBLEM SOLVING: ACT IT OUT

Some problems are best solved by the strategy of acting out the situation involved. Such an approach allows students to become active participants rather than passive spectators, and helps them to see and understand the meaning of a problem. Many routine problems of elementary algebra dealing with time, rate, and distance are especially well suited for acting out in the classroom. This not only clarifies the details of the problem, but motivates instruction as well. Let us consider several nonstandard problems for which such a strategy is desirable.

1. There are eight people at a party. Each person shakes hands with each of the other guests. How many handshakes will there be in all?

As noted in the preceding section, a difficult problem often can be solved by the strategy of first considering a smaller, related one. Thus you might begin the solution by having two students stand in front of the room and shake hands. Obviously there is only one handshake; that is, if Julie shakes hands with Amy, we would not then have Amy shake hands again with Julie. Next act out the problem with three and then with four students, and have the class actually count the number of handshakes that occur.

Almost every problem involves more than one strategy. Although the major approach here is to act out the problem, we also have used the strategy of beginning with a smaller problem, and now will utilize the strategy of making a table or list of results in order to search for a pattern. Thus actual experience provides us with this data:

Number of guests	2	3	4	5	6	7	8
Number of handshakes	1	3	6	10	15	?	?

Note the set of differences in the second row of the table.

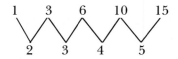

Following this pattern, the next two entries in the table would then be $15 + 6 = 21$ and $21 + 7 = 28$. Thus the solution to the original problem with 8 people is 28.

Alternatively, some students might solve the problem in a different manner. For example, note that if each of the eight guests were to stand up and walk around the room to shake hands with everyone else, then there would be $8 \times 7 = 56$ handshakes in all. But this would involve duplications; Amy shakes hands with Julie when she stands up, and then later Julie shakes hands with Amy. Every handshake between two people is made twice. Therefore the correct solution is $56 \div 2 = 28$.

Finally, note that many problems can be looked at from different points of view. In this case, consider the comparable problem of finding the total number of sides and diagonals of a polygon in which each line segment represents a handshake and each vertex represents a person. Thus we have the following visualization of handshakes for 3, 4, and 5 guests:

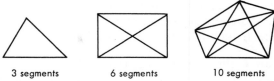

3 segments   6 segments   10 segments

As an extension, for n guests the number of handshakes would be $\dfrac{n(n + 1)}{2}$.

2. The following problem almost always elicits a wide variety of answers, even from graduate students of mathematics! It is an interesting problem to pose to a class, possibly as an assignment. After collecting the various solutions that are offered, this problem is best solved by actually acting out the situation described with play money. Try the problem; ask

a number of other people for their solutions; then satisfy yourself concerning the result by acting it out with several assistants.

A customer enters a store and purchases a pair of slippers for $5, paying with a $20 bill for the purchase. The merchant, unable to make change, asks the grocer next door for the change. The merchant then gives the customer the slippers and $15 change. After the customer has left, the grocer discovers that the $20 bill is counterfeit and demands that the shoestore owner make good for it. The shoestore owner does so, and is obligated to turn the counterfeit bill over to the FBI. How much does the shoestore owner lose by this transaction?

3.5 STRATEGY FOR PROBLEM SOLVING: MAKE A LIST, TABLE, OR CHART

We have already made use of this problem-solving strategy in several of the preceding sections. Indeed, many problems involve the use of lists, tables, and charts. Often students can be motivated to make use of this strategy by the choice of appropriate problems that stimulate their imagination and generate interest. There are a number of nonstandard problems that can serve this purpose, such as the following:

1. Sam is having a party. The first time that the doorbell rings, 1 guest enters. The second time, 3 guests enter. Thereafter, on each successive ring, a group enters that has 2 more persons than the group that had entered on the previous ring. How many guests in all will have arrived after the twentieth ring?

This problem lends itself to an initial strategy of acting out so as to clarify details. Thereafter, a table of results should indicate the pattern and lead to a solution.

RING NUMBER	NUMBER WHO ENTER	TOTAL NUMBER
1	1	1
2	3	4
3	5	9
4	7	16
5	9	25

It soon becomes evident that the total number at each stage is the square of the number of rings. That is, after the fourth ring the total number of guests will be

$$1 + 3 + 5 + 7 = 16 = 4^2$$

After the fifth ring the number of guests will be

$$1 + 3 + 5 + 7 + 9 = 25 = 5^2$$

Thus we may conclude that after the twentieth ring the total number will be

$$1 + 3 + 5 + 7 + \ldots + 39 = 20^2 \text{ or } 400$$

In general, note that the sum of the first n odd numbers can be shown to be n^2.

2. One hundred coins, all showing tails, are placed on a table of counting numbers from 1 to 100.

1	2	3	4	5	6	7	8	9	10
11	12	13	14	15	16	17	18	19	20
21	22	23	24	25	26	27	28	29	30
31	32	33	34	35	36	37	38	39	40
41	42	43	44	45	46	47	48	49	50
51	52	53	54	55	56	57	58	59	60
61	62	63	64	65	66	67	68	69	70
71	72	73	74	75	76	77	78	79	80
81	82	83	84	85	86	87	88	89	90
91	92	93	94	95	96	97	98	99	100

Now the following instructions are given in succession:

a. Change all the coins from tails to heads.
b. Change all the coins that are on even numbers from heads to tails. Thus the first row now appears as follows:

H T H T H T H T H T

c. Next change all the coins on squares that show multiples of 3. If the coin on that space shows heads, change it to tails; if it shows tails, change it to heads. The first row now looks like this:

H T T T H H H T T T

d. In a similar manner, now change all coins on squares that show multiples of 4.
e. Continue with this process up through the multiples of 100.

At the end of the process described, which numbered squares will have coins that show heads?

This problem is best solved by forming an appropriate chart that describes the situation, and then searching for a pattern. For example, consider the first 10 counting numbers and the results obtained after five changes have taken place.

	1	2	3	4	5	6	7	8	9	10
Start	T	T	T	T	T	T	T	T	T	T
First change	H	H	H	H	H	H	H	H	H	H
Second change	H	T	H	T	H	T	H	T	H	T
Third change	H	T	T	T	H	H	H	H	T	T
Fourth change	H	T	T	H	H	H	H	H	H	T
Fifth change	H	T	T	H	T	H	H	H	T	H

The reader is urged to extend the chart and continue until a pattern becomes evident. As an alternative approach, consider the number of factors for each counting number.

3. Challenge your class to guess three numbers that you are thinking of. Provide these clues:

The product of the three numbers is 72.
The sum of the three numbers is a multiple of 7.

To solve this problem, students should form a table that lists all possible sets of three factors of 72, together with the sums of these factors.

FACTORS			SUMS		FACTORS			SUMS
1	1	72	74		2	2	18	22
1	2	36	39		2	3	12	17
1	3	24	28		2	4	9	15
1	4	18	23		2	6	6	14
1	6	12	19		3	3	8	14
1	8	9	18		3	4	6	13

Students should recognize, from the table, that the solution is not unique with the information given. That is, there are three sets of factors whose sums are multiples of 7:

$$1, 3, 24 \quad \text{Sum} = 28$$
$$2, 6, \ 6 \quad \text{Sum} = 14$$
$$3, 3, \ 8 \quad \text{Sum} = 14$$

Thus their response should be that there is inadequate information to solve the problem. Now you can provide the class with an additional clue so as to determine a single solution. For example, if you indicate that the three factors are unique (no duplications), then the only solution to the problem is the numbers 1, 3, and 24. Other clues can be given to obtain different sets of factors as the solution.

3.6 CONCLUSION

It is likely that the topic of problem solving will continue to dominate discussions about the mathematics curriculum throughout the 1990s. Part of the reason for this is that mathematicians, mathematics educators, psychologists, and teachers have been unable to agree on the essential components of problem solving or the best procedures for teaching students to become good problem solvers.

Among the recommended actions suggested by the National Council of Teachers of Mathematics in *An Agenda for Action* is the following:

> Researchers and funding agencies should give priority in the 1980s to investigations into the nature of problem solving and to effective ways to develop problem solvers.

This report urges that support be given to efforts to analyze effective strategies and identify techniques for teaching problem solving. It suggests that new programs be developed to help prepare teachers for teaching problem-solving skills. To date, however, very little has been done to implement these particular recommendations.

There are many lists of problem-solving strategies available in the literature, and no universal agreement on any single collection. Among the many strategies that can be found in textbooks are the following:

Obtain an answer by trial and error.
Use an aid, model, or sketch.
Search for a pattern.
Act out the problem.
Make a list, table, or chart.
Work backwards.
Start with a guess.
Solve an equivalent but simpler problem.
Relate a new problem to a familiar one.

As the decade of the 1980s comes to a close, we find that teachers and textbooks are stressing problem solving and strategies for problem solving in the classroom. However, we still have a long way to go before all mathematics teachers recognize and understand the true nature of problem solving and its role in the mathematics curriculum at all grade levels.

EXERCISES

For each Exercise, identify one or more suitable problem-solving strategies and solve the problem.

1. How many games must be played in order to obtain a winner in a table tennis tournament of 20 players?
2. What is the units digit in the expansion of 7^{1989}?

3. What is the tens digit in the expansion of 7^{1989}?

4. What is the digit in the tens place in the expansion of 3^{1989}?

5. A number is formed by writing the counting numbers in order:

$$123456789101112131415 \ldots$$

What is the one millionth digit in this number?

6. What is the sum of the first 100 odd counting numbers? The first 100 even counting numbers?

7. Consider the matching of fractions (rational numbers) and counting numbers shown on page 63.

a. Which counting number will be matched with the fraction $\frac{15}{21}$?

b. Which counting number will be matched with the fraction $\frac{p}{q}$?

8. Consider the following charts of numbers. Place numbers in the second row so that each of the numbers in the first row appears as many times in the second row as is designated by the number placed beneath it. (Since the wording of this problem is complex, the first chart has been completed as an example.)

0	1	2	3
1	2	1	0

0	1	2	3	4

9. Repeat Exercise 8 for this chart:

0	1	2	3	4	5	6	7	8	9

10. Consider Exercises 8 and 9 and attempt to find solutions for charts that show numbers from 0 through 6, 7, and 8. Can you find a solution for a chart that shows numbers from 0 through 5?

11. Two mathematicians are each assigned a positive integer. Each knows his or her own integer, but neither one knows the other's. They are told that the product of the two integers is either 8 or 16. This is their conversation:

First mathematician:	"I do not know your number."
Second mathematician:	"I do not know your number."
First mathematician:	"Give me a hint."
Second mathematician:	"No, you give me a hint."

At this point one of the two mathematicians knows the other's number. Assuming both always tell the truth and do not guess, what is the number and who has it?

12. A penny and a quarter are on a table. You are told:

"If you tell me a true statement, I will give you one of the coins. If you tell me a false statement, you will get none."

What statement can you make that will guarantee that you will receive the quarter?

13. How many different tips are possible if you plan to use exactly three coins, and you have a penny, nickel, dime, quarter, and half-dollar available?

14. Two boats travel back and forth across a river at a constant rate of speed, without stopping. They start at the same time from opposite sides of the river and pass each other for the first time when they are 700 feet from one shore. After each makes one turn they pass again when they are 400 feet from the other shore. How wide is the river? (Draw a sketch and attempt to solve the problem without the use of algebra.)

15. Draw a network to illustrate the Koenigsberg Bridge problem. Then use Euler's conclusions about networks to show why the solution to this problem is impossible.

16. Use calculus to find the maximum volume possible for an open box formed from a 10 × 10-inch square piece of cardboard with squares of equal size cut off from each corner.

17. What is the greatest number of pieces of cake you can get with four successive slices through a cake?

18. Consider the set of counting numbers from 1 through 1989. You are to remove any two of these numbers and replace them by their difference. Continue in this manner until there is just one number left in the set. Is that number odd or even? Justify your answer.

ACTIVITIES

1. Prepare a collection of nonstandard problems that are suitable for use in grades six to nine.

2. Repeat Activity 1 for students in a senior high school mathematics class.

3. For use with gifted mathematics students, begin a collection of challenging problems such as those found on various competitive examinations.

4. Find a specific problem to illustrate each of the strategies listed in the conclusion to this chapter.

5. Begin a collection of unsolved problems, such as the three famous problems of antiquity.

6. Begin a collection of impossible problems, that is, problems that cannot be solved.

7. Complete research on the four-color problem as to the statement of the problem, the efforts made to attempt a solution in the past, and the current status of the problem.

8. Prepare a one-period lesson plan for a junior high school mathematics class that develops a specific problem-solving strategy.

9. Repeat Activity 8 for a senior high school mathematics class.

10. Prepare a one-period lesson plan suitable for a high school geometry class that provides for student discovery of patterns.

11. Begin with a 10 × 10-inch piece of cardboard, and form a box without a top by cutting off square corners of equal size and folding up the edges. Without using the calculus, experiment to find the size of the square corners cut off to obtain a box of maximum volume. Give your answer to the nearest half inch. Also compute the corresponding volume.

12. Use this program in BASIC to check your answers to Activity 11. Then modify

it to find the maximum volume possible when squares are cut off to the nearest hundredth of an inch. Compare your answers to those of Exercise 16.

```
5     REM VOLUME OF BOX
10      INPUT "DIMENSIONS ";X,Y
15      PRINT
20      PRINT "DIMENSIONS","VOLUME"
30      IF X < Y THEN C = X: GOTO 50
40      LET C = Y
50      FOR I = 1 TO .5 * C
60      IF .5 * C - I < = 0 THEN 999
70      PRINT I" X "X - 2 * I" X "Y - 2 * I,
80      LET V = I * (X - 2 * I) * (Y - 2 * I)
90      PRINT V
100     NEXT I
999     END
```

READINGS AND REFERENCES

1. Read *"An Agenda For Action: Recommendations for School Mathematics of the 1980s"*, a publication of the National Council of Teachers of Mathematics. In particular, report on the recommendations for action stated in this report concerning problem solving.

2. Read the 1980 Yearbook of the National Council of Teachers of Mathematics titled *Problem Solving in School Mathematics*. Prepare a report on Chapter 3, "Heuristics in the Classroom" by Alan H. Schoenfeld, as well as on any other three chapters of the book.

3. Read *Problem Solving: A Handbook for Teachers* by Stephen Krulik and Jesse A. Rudnick, New York: Allyn and Bacon, Inc., 1986. Report on Chapter II, "A Workable Set of Heuristics."

4. Both the *Arithmetic Teacher* and the *Mathematics Teacher*, publications of the National Council of Teachers of Mathematics, contain frequent articles in the area of problem solving at all grade levels. Review the issues of these periodicals during the past year and report on the scope of their coverage of this topic. Also, read and write a report on three selected articles on problem solving from each of these two journals.

5. Review several recently published high school algebra textbooks and report on the manner in which problem solving is treated. In particular, note any attention that may be given to strategies for problem solving.

6. Repeat Reference 5 for secondary geometry textbooks.

7. Repeat Reference 5 for several books intended for 7th and 8th grade students.

8. The 1984 Yearbook of the National Council of Teachers of Mathematics is titled *Computers in Mathematics Education*. Read Chapter 21 of that book, "Computer Methods for Problem Solving in Secondary School Mathematics" by Dwayne E. Channell and Christian R. Hirsch.

9. Read and report on Euler's work on traversable networks. One good source is Chapter 9, "Unicursal Problems" in *Mathematical Recreations and Essays* by W. W. Rouse Ball, New York: The Macmillan Company, 1960.

10. Read the following chapters in the 1982 Yearbook of the National Council of Teachers of Mathematics, *Mathematics for the Middle Grades* (5–9):
    Chapter 7: "Problem Solving for All Students" by Randall I. Charles, Robert P. Mason, and Catherine A. White.
    Chapter 20: "Wanted Dead or Alive: Problem-solving Skills" by Joyce Scalzitti.
    Chapter 21: "The Flip Side of Problem Solving" by Charles A. Reeves.

11. Read Chapters 1 through 7 of the 1983 Yearbook of the National Council of Teachers of Mathematics titled *The Agenda in Action*.

12. The 1985 Yearbook of the National Council of Teachers of Mathematics is titled *The Secondary School Mathematics Curriculum*. Read and report on Chapter 11 of that yearbook, "A Plan for Incorporating Problem Solving throughout the Advanced Algebra Curriculum" by LeRoy C. Dalton.

13. Read and report on the article by Sandra Turner, "Windowpane Patterns," in the September 1983 issue of the *Mathematics Teacher*, a publication of the National Council of Teachers of Mathematics.

14. Read *How to Develop Problem Solving Using a Calculator*, a 1981 publication of the National Council of Teachers of Mathematics by Janet Morris. Create several activity pages similar to the ones given by the author.

15. Read *Problem Solving in the Mathematics Classroom*, a 1982 publication of the National Council of Teachers of Mathematics edited by Sid Rachlin. Discuss the teaching strategies described in this publication.

Recreational and Enrichment Activities

Chapter 4

Mathematical recreations can serve as a very effective means of motivation at almost all levels of instruction and for students of varying levels of ability. The supply of such items is almost inexhaustible and can be used in many ways. Some activities can be used as an integral part of the daily lesson, others can be used to promote discovery in laboratory settings, and still others are worth introducing just for fun. The regular use of mathematical recreations throughout the year can very well serve as a means of convincing students that mathematics can be quite exciting, and can serve to transfer this enthusiasm to other mathematical topics in the traditional curriculum.

Within this chapter are presented a variety of mathematical games, puzzles, and enrichment topics. For each of these categories, only a relatively small number of illustrative examples are given. Hopefully the reader will be stimulated to search the literature and follow the suggested bibliographical leads to begin a collection of many more of these recreational activities.

4.1 CLASSROOM GAMES

Mathematical games can be used effectively for a variety of purposes. They can be used solely as recreational devices to motivate a class and generate interest. As such, they supply a good source of material for use during the last few minutes of a period, the day before a holiday, and in similar circumstances. Their use as part of the program for a mathematics club can almost always be counted upon to interest student members.

In addition to purely recreational aspects of games, other objectives can be attained through their use. Many mathematical games can be used to lead students to formulation and testing of hypotheses as they strive to discover a winning strategy. The development of such modes of thinking has long been recognized as a worthwhile outcome of the teaching of mathematics, and is truly the essence of problem solving.

Finally, mathematical games can be used as an effective way to develop certain basic concepts and skills. Arithmetic and geometric skills, as well as ability to visualize in two and three dimensions, are just a few of the many mathematical items that can be approached through the use of appropriate games.

The following collection is only a representative sample of suitable games for the mathematics classroom. The reader is referred to the list of readings on page 101 for further sources of information on this topic.

The Game of 50

The Game of 50 is designed for two players. An effective way to introduce the game is to announce that you are the world champion at this game, and that you are willing to have a student challenge you. Then allow one or two students to compete with you during the last few minutes of the period each day for several days. Continue until someone in the class notes the pattern you use for winning and is able to win the game.

Rules for playing The game is played using the numbers 1, 2, 3, 4, 5, and 6. The two players alternate in selecting numbers, and the first to reach 50 wins. As each new number is selected, it is added to the sum of the previously selected numbers. For example, if the student goes first and selects 3, the teacher might then select 6, to give a sum of 9. If the student then selects 5, the total is 14, and it becomes the teacher's turn to go. The game continues in this manner until one player becomes the winner by reaching 50.

Strategy Actually this game is an excellent way to illustrate the problem-solving strategy of working backwards. Analysis of the game shows that you can always reach 50 if you first reach 43. (Regardless of what number your opponent selects then, you can choose a number to obtain 50.) Working backward, you can reach 43 if you can get to 36. Continuing in this way, the following "winning numbers" are obtained:

$$1, 8, 15, 22, 29, 36, 43, 50$$

Thus the strategy for winning is to go first and begin with 1. Thereafter, select the complement of your opponent's number relative to 7. That is, if your opponent selects 4, you choose 3; if your opponent selects 2, you choose 5; and so forth. If your opponent goes first and does not know the rules for winning, choose your numbers so as to reach one of the winning numbers as soon as you can.

Extensions   Many variations of this game are possible. A similar game is played by using a set of 16 cards consisting of the four aces, four 2's, four 3's, and four 4's.

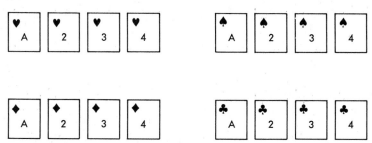

Players alternate selecting one card at a time from the pile of 16 cards, face up, without replacements. As before, cumulative sums are kept. The winner is the first person to select a card that brings the total to exactly 22, or forces the opposing player to go over 22. The set of winning numbers for this game is

$$2, 7, 12, 17, 22$$

The strategy for winning is to go first and begin with 2. Thereafter, select the complement of your opponent's number relative to 5. That is, if your opponent picks 4, you choose 1, and so forth. However, in this case this is not a foolproof strategy because the number of cards is limited. Suppose that your opponent repeatedly chooses 3. This would force you to repeatedly choose 2 and you would run out of 2's prior to reaching the objective number of 22. Can you consider alternative strategies for such a situation?

The Game of Sprouts

Sprouts is a game for two that is most effectively presented by dividing the class into pairs to play against one another. The one who wins two out of three games can then be declared the winner, and the winners paired against one another again until a class champion emerges.

Rules for playing   The game starts with two points, labeled *A* and *B* in the figure, that are called *spots*. Each player takes a turn drawing an arc from one spot to another, or back to the same spot, and then places a new spot on the arc drawn. For example, here are two possible moves that the first player might make:

The two basic rules of play are that no arc may cross itself or pass through another arc or spot, and that no spot may have more than three arcs from that point. The winner is the last person who is able to draw an arc. Here is an example of a game played; a circle around a point indicates that there are three arcs at the point and thus the point is no longer in play:

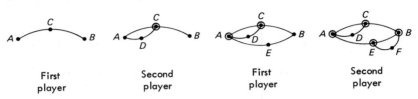

| First player | Second player | First player | Second player |

In the game illustrated, the second player wins. The first player cannot draw an arc from D to F because it would pass through another arc. Also, there is no way to draw arcs from D to itself (or from E to itself) because then there would be four arcs at that point.

Strategy Note that each spot becomes a dead spot once it is used three times as the endpoint of an arc. Therefore, the game begins with six possible "lives." The first player uses up two lives, but adds one life by adding a spot to the new arc drawn. Therefore, after the first play the game has five lives left. In a similar manner, there are four lives left after the second move, and only one life left after the fifth move. With only one arc available, the game must be at an end. Therefore, the game has a maximum of five possible moves, a small enough number for a student to be able to draw the set of all possible games.

Here is an example of a game that utilizes the maximum number of moves, resulting in a win for the first player:

| First player | Second player | First player | Second player | First player |

Extensions The game can be played using any number of spots initially. Consider playing the game using three spots, in which case there will be a maximum of eight possible moves. The maximum number of moves possible in four-spot Sprouts is 11.

The Game of Nim

Although Nim may be used purely as recreation at any time, it is best presented after the student has some knowledge of the binary system of numeration, thus providing an interesting application of that topic. The game is played by two students at a time and is probably best introduced by having the teacher challenge a student to a match.

Rules for playing Many versions of the game can be played. Probably the easiest one to start with consists of three piles of chips. Pile A contains 3 chips, pile B contains 4 chips, and pile C contains 5 chips. At your turn, you select one pile of chips and remove as many chips as you wish from that one pile. You must, however, remove at least 1 chip. Players alternate, and the player who picks up the last chip on the board is the winner.

Strategy Students may develop informal strategies for winning. The formal winning strategy is quite complex and consists of writing the number of chips in each pile in binary notation. In order to win, one must be certain that the sum of the digits in each binary place is even after the move. To illustrate this principle, here is an example of a game where the first player wins:

1.	A	XXX	In binary notation:	11	First player takes 2
	B	XXXX		100	chips from pile A,
	C	XXXXX		101	as shown next.
2.	A	X	In binary notation:	1	Second player
	B	XXXX		100	takes 3 chips from
	C	XXXXX		101	pile C.
3.	A	X	In binary notation:	1	First player takes 1
	B	XXXX		100	chip from pile B.
	C	XX		10	
4.	A	X	In binary notation:	1	Second player
	B	XXX		11	takes both chips in
	C	XX		10	pile C.
5.	A	X	In binary notation:	1	First player takes 2
	B	XXX		11	chips from pile B.
6.	A	X	In binary notation:	1	Second player
	B	X		1	takes 1 of the chips, and the first player wins on the seventh move by taking the remaining chip.

Note in the above game that the first player always made a move that left the second player with an array of chips whose number in binary notation had a even number of 1's in each column.

Extensions The game can be played using the rule that the one who is forced to take the last chip is the loser. It can also be played with an indefinite number of chips in each of the original three piles, although the strategy for winning remains the same.

The Game of Tac Tix

Tac Tix, a game for two, is a variation of the game of Nim, which was invented by Piet Hein of Denmark in recent years. It is an interesting one to present in that it has not yet been completely analyzed.

Rules for playing Arrange 16 coins or chips, as in the figure. These are numbered here for ease of reference.

①	②	③	④
⑤	⑥	⑦	⑧
⑨	⑩	⑪	⑫
⑬	⑭	⑮	⑯

Players alternate removing any number of chips from any single row or column. However, as an additional constraint, only adjacent chips may be removed. For example, if player A removes chips 14 and 15 on his first move, player B may not take 13 and 16 in one move. The player who is forced to take the last chip is the loser.

Strategy It is necessary to play this game so that the one who takes the last chip loses because otherwise the first player would always win. On a 3 × 3 board, using 9 counters and following the rules stated above, the first player can win by taking the center chip, or a corner chip, or all of a central row or column. There is no known strategy for winning the game with 16 chips as described above.

Extensions For further reading on this game, see *Scientific American Book of Mathematical Puzzles and Diversions*, Martin Gardner, ed. (New York: Simon and Schuster, 1964, pp. 157–160). In this account Gardner suggests that one can best gain an introduction to this game by solving specific problems. Thus, it seems worthwhile to present Tac Tix problems for students to solve before actually playing the game with an opponent. Gardner presents these two problems suggested by the inventor, Piet Hein. In each case you are to find a move that will guarantee a win.

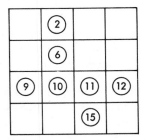

The Game of Hit-and-Run

Hit-and-Run is another game for two that requires some careful thinking on the part of the players. It is most conveniently played on graph paper, and is similar to a number of commercial games currently on the market.

Rules for playing The game starts with a square grid. Players alternate coloring one line segment at a time in an effort to build a bridge from one side of the grid to the other. For example, the figure shows a winning path for player *A* that goes from *A*'s chosen side of the board to the opposite side. The subscripts indicate the order in which the moves were made. (Note that opponents' paths may cross one another.)

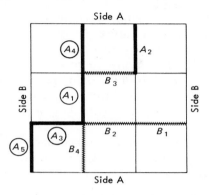

Strategy For a 2 x 2 grid, the first player can always win by following the indicated moves.

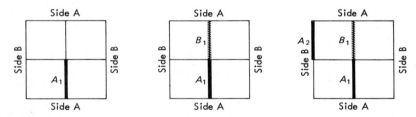

At this point *A* has two possible paths to complete, and therefore must win regardless of where *B* moves.

For a 3 x 3 grid, the first player has an advantage if the first move made is in either one of the positions shown. Verify this by playing several games.

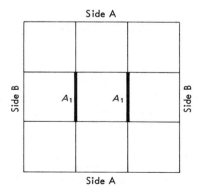

Extensions Play the game on a 4 x 4 grid. At the moment it is not known whether there is a winning strategy for this size game nor whether there is any advantage to going first.

4.2 PUZZLE PASTIMES

Most students of mathematics enjoy working with puzzles. Although these are most often recreational in nature, nevertheless there are many other worthwhile outcomes that accrue from the use of suitable puzzles and problems. Appropriate recreational items can be found to stimulate intellectual curiosity, to develop abilities in space perception, to promote discovery, and to develop modes of thinking. The major purpose for their use in the classroom, however, is to stimulate interest in the further study of mathematics.

There are a variety of effective ways that mathematical recreations can be used in the classroom. The reader should be able to add to the following list of suggestions:

1. Have one section of the bulletin board or chalkboard entitled "Puzzle of the Week." Each week place a new puzzle or set of puzzles in this spot. Encourage students to submit answers in writing. At the end of the week post a list of the names of all students who submitted correct solutions. Possibly award a prize for the first correct answer (such as an excuse from homework for one night), or a prize to the student who has the greatest number of correct solutions within each marking period.
2. Place a collection of puzzles on 4 x 6-inch index cards. Any student who completes a classroom assignment or test early is allowed to come up and select a card to use while waiting for the rest of the class to finish their work.
3. Devote the last five minutes of each period to a mathematical recreation.
4. Devote the last half of the period each Friday to mathematical recreations.
5. Use the last period before a vacation for recreational activities.
6. On occasion include an interesting puzzle as part of the regular assignment.

The supply of interesting puzzles is almost endless, and the reader is referred to the Readings on page 101 for additional sources of such items. The following list is only representative of the type of puzzle that appears to be of interest to secondary school mathematics students. Some of these appear again in the Exercises.

1. How can you cook an egg for exactly 15 minutes, if all you have is a 7-minute hourglass and an 11-minute hourglass?
2. How can a 24-quart can of water be divided evenly among three unmarked cans whose capacities are 5, 11, and 13 quarts?
3. Nine coins are in a bag. They all look alike, but one is counterfeit. It weighs less than the others. Use a balance scale and find the fake coin in exactly two weighings.
4. There are 12 coins, of which one is counterfeit, weighing less than the others. Use a balance scale and find the fake coin in exactly three weighings.

5. Use the digits 2, 3, 4, 5, 6, 7, 8, 9, and 10. Place exactly one of these in each position in the figure so that the sum for each row, column, and diagonal is 18.

6. Arrange the numerals 1 through 8 in the figure so that no two consecutive integers touch at a side or on a corner.

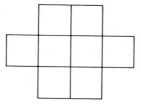

7. Arrange eight coins as in this figure:

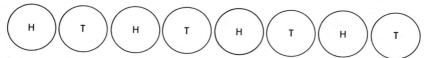

Moving two adjacent coins at a time, try to obtain the following arrangement:

8. Draw a square, number eight pieces of paper, and arrange them as shown in the figure on the left. By sliding only one piece at a time into an open square, arrange the pieces as shown in the figure on the right.

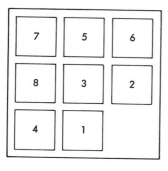
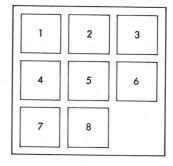

9. Arrange three piles of toothpicks, chips, or other similar objects so as to have 6 items in pile A, 7 items in pile B, and 11 items in pile C. In exactly three moves you are to attempt to obtain 8 items in each pile. The rules for movement are that you may only move to a pile as many items as are already there, and all items moved must come from a single other pile.

10. Arrange five coins as in this figure:

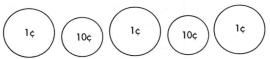

Try to obtain the following arrangement by moving two adjacent coins at a time, but each pair of coins moved must consist of a penny and a dime and must not be interchanged during the move.

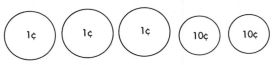

11. Eleven toothpicks are arranged as shown to give five triangles.

    (a) Remove one toothpick to show four triangles.
    (b) Remove two toothpicks to show four triangles.
    (c) Remove two toothpicks to show three triangles.
    (d) Remove three toothpicks to show three triangles.

12. A coin is in a "cup" formed by four matchsticks. Try to get the coin out of the cup by moving only two matchsticks to form a congruent cup but in a new position.

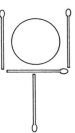

    Some puzzles of a geometric nature simply require great amounts of patience and ingenuity.

13. Loosely tie together the hands of two people with one string looped around the other, as shown. Now get them apart without untying or cutting the string. It can be done!

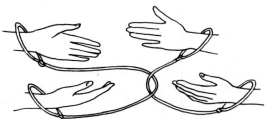

14. Cut a tag like this one. Pass a string through the hole, under the narrow center strip, out again, and back through the hole. Knot the ends. The trick is to get the string off the tag without cutting the string, or tearing the tag, or passing the string back through the hole.

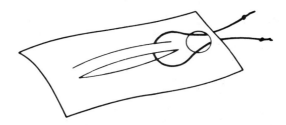

Quite a different type of puzzle that can be both amusing and useful in the mathematics classroom deals with letter and word arrangements and codes. Some are straightforward and reasonably simple, whereas others can tax the most capable. Students enjoy making as well as solving these problems.

15. Can you break these alphametics?

Addition:		Subtraction:	
	O N E		F I V E
	T W O		F O U R
	F I V E		O N E
	E I G H T		

16. A student at college sent the following message home.

```
    W I . R E
    M O . R E
  M O N . E Y
```

If each letter represents a unique number, how much should be sent?

Many of these puzzle pastimes lend themselves to a series of problems or activities that can be presented to the student in the form of laboratory worksheets or activity cards. Here is one example. Many others can be found in the following chapters.

17. Start with the A on top. Move down to the left or right one letter at a time. The path shown spells out the word ALGEBRA.

```
          A
         L L
        G G G
       E E E E
      B B B B B
     R R R R R R
    A A A A A A A
```

How many different paths are possible? Do they all spell ALGEBRA?

4.3 MATHEMATICS IN FAMILIAR GAMES

In addition to the recreational aspects of games, many games can be used effectively in the classroom to motivate the study of specific mathematical topics. In this section some mathematical applications are illustrated for the familiar games of tic-tac-toe and pool. The use of mathematics in other games, such as checkers, chess, poker, and roulette, will be discussed in subsequent chapters.

Tic-tac-toe

A careful study of two- and three-dimensional tic-tac-toe can produce some interesting applications to counting and coordinates that can be both useful and motivating in the mathematics classroom. Here are some suggested activities.

Set up the grid on a pair of axes, and play the game with the class using ordered pairs of numbers to locate moves.

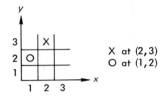

X at (2,3)
O at (1,2)

Present partially completed games and have the students suggest subsequent moves. For example, if these moves have been made, where should O move next?

O's at (1, 3), (2, 2), (1, 1)
X's at (3, 3), (1, 2), (3, 2)

Together there are eight possible winning arrangements in the game of tic-tac-toe. Have students give the number pair needed to make these winning arrangements.

(3, 1), (3, 3), _?_
(1, 3), (2, 2), _?_
(2, 2), (3, 2), _?_

Students can then practice giving the correct equations for different possible winning arrangements.

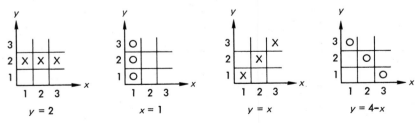

$y = 2$ $x = 1$ $y = x$ $y = 4-x$

Applications from three-dimensional tic-tac-toe follow in much the same way only at a decidedly higher level of difficulty. Some suggested classroom activities are described here.

Set up the grid on three mutually perpendicular axes and practice locating positions using number triples in the form (x, y, z). Have students describe some four-in-a-row wins. It is a nontrivial exercise simply to count the 76 possible winning arrangements on a 4 x 4 x 4 grid.

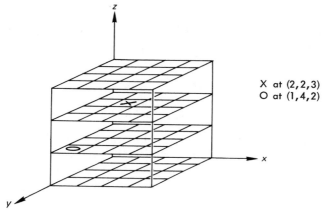

X at (2,2,3)
O at (1,4,2)

Students can give the number triple needed to complete a winning arrangement such as in these examples.

(2, 3, 1), (2, 3, 2), (2, 3, 3), _?_
(1, 1, 1), (2, 2, 2), (3, 3, 3), _?_
(4, 4, 4), (4, 3, 3), (4, 2, 2), _?_

An interesting question would be to find the seven winning arrangements containing a given position such as (3, 3, 3).

_?_, _?_, (3, 3, 3), _?_

If the idea of counting possible winning arrangements is extended to each of the 64 positions on the grid, the player would have a good idea as to the best moves to make initially in playing the game.

Pool

Some interesting applications of geometry can be found in the game of pool. For example, consider these properties and their possible use when teaching about angles.

The angle of rebound off a cushion is the same as the angle of approach.

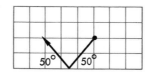

50°   50°

The angles of rebound off opposite cushions are the same. The new direction of the ball is parallel to its initial direction because the alternate interior angles are congruent.

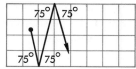

The angles of rebound off adjacent cushions are complementary. Here again the new direction of the ball is parallel to its initial direction.

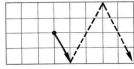

A variety of simple exercises illustrating these properties can be assigned using graph paper to establish the shape and dimension of the table and to facilitate the easy construction of congruent angles. For example, describe the path of the ball when shot in the direction shown.

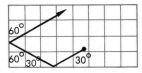

The Peg Game

This game is usually played by moving pegs along a row of holes in a block of wood. Black and white pegs move in opposite directions. The object of the game is to interchange the positions of the black and white pegs in as few moves as possible. The rules of the games are as follows:

1. You may move only one peg at a time.
2. Black pegs move only to the right and white pegs move only to the left.
3. You may move a peg into an empty hole, or jump over a single peg into an empty hole.

For a version of the game that can be conveniently played by your students, use dimes and pennies to represent the black and white pegs and a strip of squared paper for the board. As a first example, consider the figure below with two dimes at the left and two pennies at the right, and an empty square in the middle.

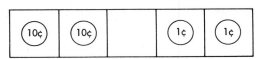

Students should be encouraged to play the game and search for a solution by trial-and-error methods initially. Then suggest that they search for a pattern that would allow them to predict the minimum number of moves necessary for any number of pairs of coins. To do so, they should consider the problem-solving strategy of solving equivalent, but simpler, problems first. That is, they might consider a game with one dime, one penny, and one empty square in the middle. It should be easy to determine that this game can be completed in just three moves.

By solving the initial problem given here, students will find that eight moves are necessary for two pairs of coins. Thereafter have them consider games with three and four pairs of coins. In order to discover a pattern in the sequence of moves needed, encourage students to record systematically all moves made. One choice is to use D for a dime moving to the right and P for a penny moving to the left. At any given stage, only one move is possible for each coin. The solutions given here all start with the dime moving to the right. (Comparable solutions with the penny moving to the left first can be formed by simply interchanging all D's and P's.)

NUMBER OF PAIRS	SEQUENCES OF MOVES	NUMBER OF MOVES	NUMBER OF JUMPS
1	DPD	3	?
2	DPPDDPPD	8	?
3	DPPDDDPPPDDDPPD	15	?
4	DPPDDDPPPPDDDDPPPPDDDPPD	24	?

To find a rule that relates the number of moves to the number of pairs of coins, consider the results in the table and find first and second differences as follows. (See page 170.)

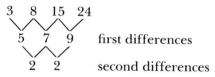

first differences

second differences

If the second differences are constant, then the relationship between the number of moves, m, and the number of pairs, p, is quadratic and of the form

$$m = ap^2 + bp + c$$

To solve for a, b, and c, use data from the table to form these equations:

$$p = 1, m = 3: \quad 3 = a + b + c$$
$$p = 2, m = 8: \quad 8 = 4a + 2b + c$$
$$p = 3, m = 15: \quad 15 = 9a + 3b + c$$

Solve these three equations to find $a = 1$, $b = 2$, $c = 0$. Thus:

$$m = p^2 + 2p = p(p + 2)$$

So, for five pairs of coins ($p = 5$), the number of moves will be $5(5 + 2) = 35$. For ten pairs of coins, the number of moves will be 120. And for n pairs of coins, the minimum number of moves required will be $n(n + 2)$.

This generalization for the minimum number of moves required does not, however, give the particular sequence of moves right by the dime and left by the penny. Can you discover the pattern from the first four sequences of moves given in the table? Predict the sequence of D's and P's required for the 35 moves needed for five pairs of coins. Then test your prediction by actually moving the coins using this sequence.

As an interesting extension of this game, count the number of jumps made in each case, look for a pattern, and then generalize your solution.

4.4 ENRICHMENT TOPICS

Many topics in mathematics are of interest in themselves, without regard for any particular application to the physical world or to other branches of mathematics. These are classified as enrichment topics and are most suitable for presentation in the mathematics class at a variety of grade levels.

These topics are appropriate to use as tangential to routine classroom activities and as such are effective means of motivation. Many of these can be expanded to form the basis of interesting laboratory lessons, whereas others lend themselves quite well to bulletin board displays and student projects.

As is true throughout this chapter, the items presented herein are merely representative of the large number of enrichment topics available for classroom use.

Polyominoes

A polyomino is merely a set of squares connected along their edges. The simplest form is a single square, called a *monomino*. Two connected squares are called a *domino*, three squares are called a *triomino*, and four connected squares are called a *tetromino*. In this enrichment topic we are concerned with the total number of possible arrangements of such figures that are not congruent to one another.

Classroom procedures Supply students with graph paper since this is the most convenient way of drawing and studying polyominoes. Together with the class, demonstrate the following figures and arrangements:

There is only one type of monomino and one domino.

There are two possible arrangements of triominoes that use three squares.

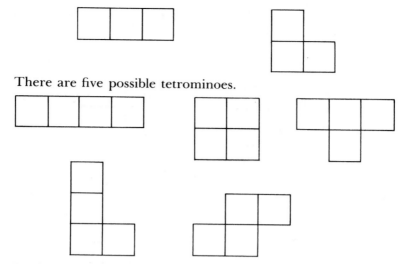

There are five possible tetrominoes.

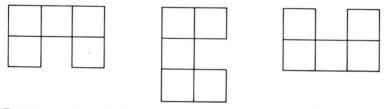

Having offered a class the preceding exposition, ask them to find all the possible pentominoes, figures formed by five connected squares. Caution them not to include any that are congruent to one another. For example, each of the following consists of just a single arrangement.

Extensions An obvious extension is to have students search for all possible hexominoes, arrangements of six squares. However, this is a tedious and difficult project inasmuch as there are 35 different hexominoes. This could be a class project where each newly discovered hexomino is placed on the board until all 35 are found.

Another interesting, but difficult extension is to cut out all 12 possible pentominoes and try to arrange them to form a rectangle that is 5 units by 12 units in dimension. This is the basis for a puzzle called *Hex*. Yet another interesting extension is to determine which of the 12 pentominoes can be folded to form a box without a top. A problem such as the following can also be used as a suitable classroom activity after a discussion of polyominoes.

If the pattern below were assembled to form a cube, it would spell MATH around four of its faces.

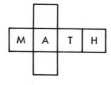

Letter the two patterns below so that they too will spell MATH the same way when assembled. Check your answers by actually forming the cubes.

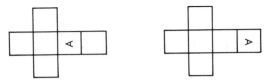

Letter these patterns so that each spells MATH when assembled. Cut out the patterns to check your answers.

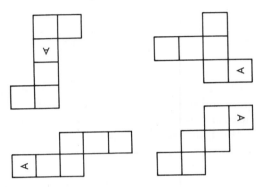

Tessellations

It is apparent to most that squares, rectangles, equilateral triangles, and regular hexagons can be used to tile or tessellate a plane. But students can explore other polygons as well to see if they can be used to form a tessellation.

Classroom procedures As an interesting classroom activity, have students cut out of paper some congruent quadrilaterals with no two sides parallel. Then see if they can arrange them to cover the plane without spaces and without overlapping. It will come as a surprise to many that any quadrilateral, convex or concave, can be used for a tessellation. Two are illustrated here. Note that the four angles at each intersection point have the same sizes as the four angles of an individual quadrilateral. Hence, they have a sum of 360° and completely cover the plane about that point.

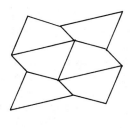
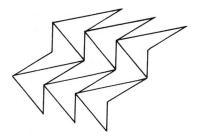

Another way for students to explore tessellations in the classroom is with graph paper or dot paper. Here is one tessellation drawn on graph paper using a hexomino of six squares as the basic shape. Try drawing some other tessellations using the same hexomino.

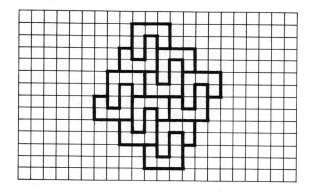

Many artists have used the idea of tessellations in their work. Perhaps the most famous is Maurits Escher (1898–1972). He created elaborate figures that would fit together to form tessellations such as this one. Encourage students to try to create similar tessellations of their own based on equilateral triangles, squares, and regular hexagons.

Extensions   A regular tessellation is made from congruent regular polygons joined side-to-side. Equilateral triangles, squares, and regular hexagons are the only regular polygons that can be used by themselves for a tessellation. A semiregular tessellation uses two or more different regular polygons with sides of the same length in such a way that all vertices are identical. The numerical notation shown for these semiregular tessellations represents the regular polygon arrangement about each vertex.

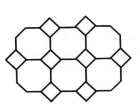

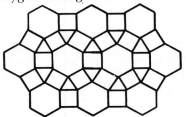

Square, octagon, octagon
(4.8.8)

Triangle, square, hexagon, square
(3.4.6.4)

There are eight semiregular tessellations in all. Two are shown above; two more (4.6.12 and 3.12.12) contain regular dodecagons. See if your students can draw these remaining two pairs using equilateral triangles, squares, and regular hexagons, all with sides of the same length:

1. 3.3.3.4.4 and 3.3.4.3.4
2. 3.6.3.6 and 3.3.3.3.6

Möbius Strips

Discovered by the German mathematician August Möbius, the Möbius strip is a fascinating item that lends itself very well to a worthwhile enrichment topic that can be presented in the form of a laboratory exercise.

Classroom procedures   Have each student begin with a strip of paper about 20 inches long and 4 inches wide for ease of handling. Mark one end A and the back of the opposite end B as in the figure.

Turn over one end of the paper so as to form a half-twist.

Now join the ends so as to form the figure known as the Möbius strip.

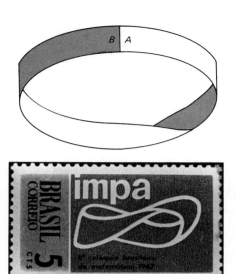

The Möbius strip is a one-sided figure. Start at *A* and draw a line down the middle of the strip. You will ultimately reach *B* without having to cross an edge, even though *B* was on the opposite side of the strip after making the initial half-twist. Now cut the figure down the middle. Instead of two figures as expected, you will end up with one band!

Extensions   Many extensions of this laboratory enrichment topic are possible. A few possibilities are listed.

1. Cut the newly formed figure down the middle again to see what type of figure is obtained, but first ask the class to predict the outcome.
2. Form a new Möbius strip, but this time cut along a line that is approximately one-third of the way across the band.
3. Form a band with two half-twists. Cut it down the center and discover the resulting figure.
4. Repeat the preceding extension, but begin with a band that has three half-twists.

Curve Stitching

Curve stitching is an enrichment topic of particular interest because it seems to capture the attention of students of varying levels of ability. It is an effective topic to use before a holiday, and can serve as the basis for very dramatic bulletin board displays.

Classroom procedures   Before allowing students to complete designs of their own, it is well to first have everyone construct one together under the teacher's guidance. A basic one with which to start begins by drawing an angle and marking off the same number of equally spaced units on each side of the angle. In the figure, 12 units are located and marked as shown.

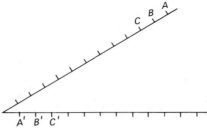

Next connect point A to point A', B to B', C to C', and so on. The line segments drawn will appear to form a curve called a *parabola*.

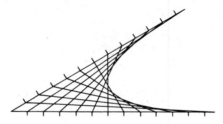

Many different variations are possible merely by changing the angle between the line segments, the distance between points, and by combining several curves. Here is an example of one possible variation.

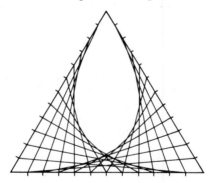

This enrichment topic is entitled "curve stitching" because the figures can be made by using colored thread and stitching through cardboard. Push up through each point from the back, and then stitch between points in the same manner as you would draw line segments.

Extensions   It is worthwhile to prepare a school exhibit consisting of a variety of curve-stitching designs made by individual students. A contest can be held with viewers asked to vote for the most attractive as well as the most original design created.

The figure shown is a far more difficult design to complete and may be suggested to some of the more industrious students in a class. The figure consists of a 24-sided polygon together with all its diagonals, giving the illusion of a series of concentric circles.

Tangrams

One of the oldest known puzzles is the ancient Chinese game of tangrams. Having amused and challenged people for thousands of years, it is certain to capture the interest of many students as well.

Classroom procedures Draw this figure on a square piece of paper or cardboard. The points are lettered here for construction purposes only and are not needed in the activity itself. First locate points M and N, the midpoints of sides AB and AD. Draw MN. Then draw diagonal BD and part of the other diagonal PC as shown. Finally, draw PQ parallel to AB and NR parallel to PC. Cutting the figure along the lines drawn gives 7 separate pieces consisting of 5 triangles, 1 square, and 1 parallelogram.

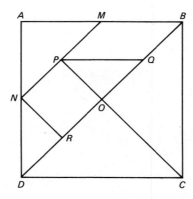

The tangram puzzle consists of rearranging these 7 pieces into some other figure. For fun at the lower grades, this figure could be a house or a dog or some such thing. A particularly appropriate adaptation for the junior high is to arrange them into convex polygons. There are 13 different convex polygons possible including 1 triangle, 6 quadrilaterals, 2 pentagons, and 4 hexagons. One of each is shown here. Note that different arrangements of the pieces within each polygon are possible.

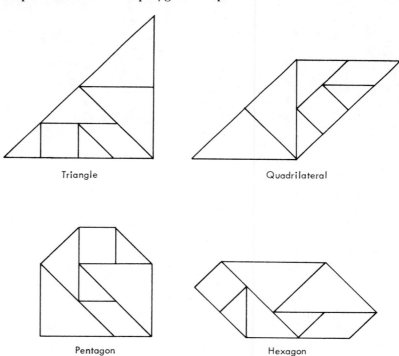

Triangle

Quadrilateral

Pentagon

Hexagon

Challenge your students to find all 13 solutions. The complete set makes an attractive bulletin board display.

Tetratetraflexagons

This amazing little geometric model appears to have just two sides, front and back. But by flexing the model, both sides can be made to disappear. It is called a *tetratetraflexagon* because it can be flexed to show four (tetra) different sides or faces and because it has a four (tetra)-edged rectangular shape.

Constructing and assembling the tetratetraflexagon requires care, patience, and a certain degree of skill, but the effort will be more than repaid by the novelty of the completed model. Students will want their own model to keep.

Classroom procedures

1. Reproduce the pattern and markings as shown so that each student can get a copy. Starting with small 2-inch squares, the initial pattern will be 6 x 8. The size may be changed, but it is advisable to avoid extremely large or small models.

		DIRECTIONS: Fold vertically down the center so the two halves are back to back. Carefully separate the two parts at the fold.	
TETRA- **TETRA-FLEXAGON**			
	A new side should appear. Flex it again and you'll find still another side. Remember, don't tear the paper.		You may think this has just two sides, front and back. If you do, you're wrong!
SIDE 1	When you flex this paper in just the right way, you can make the writing disappear.	Now have fun trying to flex them back! **SIDE 2**	

2. Cut out the pattern. Then repeatedly fold it back and forth along each of the three vertical lines. With a knife or a razor blade cut around three sides of the two center squares as shown. Be sure not to completely detach them from the rest.

3. Closely follow these assembling instructions.

Tetra-tetra-	flexagon	

Tetra-tetra-	flexagon	

Tetra-tetra-	flexagon	

Tetra-tetra-	flexagon	

Position with word tetratetraflexagon in upper left-hand corner.

Fold center flap behind left hand column and also fold back right hand column.

Fold back right hand column a second time.

Fold end of flap over on top and tape *only* on squares shown.

4. Follow directions on the tetratetraflexagon for flexing. Be careful that you do not force a fold or tear the paper. By flexing the model twice, both faces with the writing should disappear.

Extensions If this flexagon proves interesting, try assembling and flexing a hexahexaflexagon—six faces formed from a six-sided polygon. A pattern is given in the first reference to Martin Gardner found in the Readings.

Tower of Hanoi

The Tower of Hanoi is a famous mathematical problem that students enjoy trying to solve. Although some students can easily construct display-size models for use in class, it can be played using three objects of different size, such as a quarter, nickel, and penny.

Explain the rules for playing. The Tower of Hanoi puzzle consists of three disks, or pegs, of decreasing size, with the largest item on the bottom. The object of the game is to transfer the disks to one of the other pegs, following these conditions:

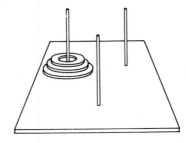

1. Move only one disk at a time.
2. No disk may be placed on top of one smaller than itself.
3. Use the fewest possible moves.

After explaining the rules of the game, students should be allowed to try to complete the game using three coins and three possible positions, as in the next figure. The three coins must be moved from position *A* to either *B* or *C*, using the rules given above. With three objects, only 7 moves are needed.

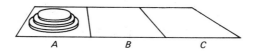

After students are successful at completing the game in 7 moves, let them attempt the game using four objects. This can be tried using a quarter, a nickel, a penny, and a dime. For four objects, and using the same rules, 15 steps are necessary. For five objects, 31 steps are needed. In general, for n objects, $2^n - 1$ steps are needed.

An ancient Hindu legend states that Brahma placed 64 disks of gold in the temple at Benares and called this the tower of Brahma. The priests were told to work continuously to transfer the disks from one pile to another in accordance with the rules set forth earlier. The legend states that the world would vanish when the last move was made. The minimum number of moves to complete this task would be $2^{64} - 1$. Ask students to estimate how many moves this would be and how long it would take at the rate of one move per second. It is interesting to note that

$$2^{64} - 1 = 18,446,744,073,709,551,615$$

The world seems safe from destruction!

Suggested Topics

There is an almost endless supply of topics that are suitable for enrichment of mathematics classes at various grade levels as well as at different levels of difficulty. The topics that follow are highly recommended for use in junior and senior high school mathematics classes.

Ancient Egyptian Mathematics	The Nine-Point Circle
Card Tricks	Nomographs
Conic Sections	The Number e
Diophantine Equations	The Number Pi
Divisibility of Numbers	Paper Folding
Euler's Formula	Paradoxes
The Euler Line	Pascal's Theorem
Fermat's Last Theorem	Perfect Numbers
Fibonacci Numbers	Prime Numbers
Finite Differences	Problems of Antiquity
Flatland	Projective Geometry
Flexagons	Pythagorean Triples
The Four-Color Problem	Regular Polygons
The Fourth Dimension	Regular Polyhedrons
Game Theory	Relativity
Geometry of Soap Bubbles	Shortcuts in Computation
The Golden Section	Tessellations
Koenigsberg Bridge Problem	Topology
Linear Programming	Trachtenberg System of Computation
Linkages	Transfinite Numbers
Mascheroni's Construction	Unsolved Problems
Mathematics and Music	Vectors
Napier's Rods	Zeno's Paradoxes

Many of the preceding topics can be assigned to individual students for research and for class projects and reports. Almost all are exceptionally well suited for mathematics club activities.

EXERCISES

1. How can a 24-quart can of water be divided evenly among three unmarked cans whose capacities are 5, 11, and 13 quarts?

2. Using two containers that hold 4 and 9 liters each, how can you obtain exactly 6 liters of water?

3. Nine coins are in a bag. They all look alike, but one is counterfeit; it weighs less than the others. Explain how to use a balance scale to find the fake coin in exactly two weighings.

4. Repeat Exercise 3 for 12 coins, using three weighings.

5. As a very challenging problem, repeat Exercise 4 but this time you do not know whether the counterfeit coin is lighter or heavier than the others.

6. A student at college sent the following message home.

$$
\begin{array}{r}
\text{W I . R E} \\
\text{M O . R E} \\
\hline
\text{M O N . E Y}
\end{array}
$$

If each letter represents a unique number, how much should be sent?

7. Using a 7-minute and an 11-minute hourglass, explain how to boil an egg for exactly 15 minutes?

8. There are three pentominoes shown on page 89. Draw sketches to illustrate all 12 possible pentominoes, and indicate which of these can be folded so as to form a box without a top.

9. There are 35 possible hexominoes. Draw all of those that contain exactly four monominoes in a row.

10. Consider a checkerboard with 64 squares, and a set of dominoes each of which can cover exactly two squares. Can you use the dominoes to cover all of the squares on the board with the exception of two in diagonally opposite corners?

11. There are three boxes that contain, respectively, two dimes, two pennies, and one dime and one penny. Each box is labelled *incorrectly*. Tell how you can take one coin out of one of the boxes (without looking) and use that choice to determine the contents of all three boxes.

12. A club consists of either liars or truthtellers. Suppose you meet three officers of the club and ask each if they are liars or truthtellers, and the conversation is as follows:
President: mumbles
Vice-President: "He said he was a truthteller; he is and so am I."
Treasurer: "That's not true. The President is a liar and I am a truthteller."
Who is lying and who is telling the truth?

13. Draw a pentagon that can be used in a tessellation. Then draw one that will not tessellate the plane.

14. Draw the semiregular tessellations represented by the notations 3.3.3.4.4 and 3.3.4.3.4.

15. A cube can be used to tessellate or fill space without gaps. Describe two other prisms and two pyramids that can also be used for a tessellation of space.

ACTIVITIES

1. Prepare a collection of 25 mathematical puzzles, placing each on an index card with answers on the reverse side.

2. Select two of the enrichment topics listed on page 99 and prepare a 15 minute oral report on the topic, using some aid in addition to the chalkboard for your presentation.

3. Revise the rules given on page 74 for "The Game of 50" to have 100 as the goal. Develop a strategy for winning, and then play the game with members of the class to demonstrate the winning strategy.

4. Prepare a bulletin board display of curve stitching art.

5. Attempt to reproduce the figure shown on page 95 of a 24-sided polygon together with all its diagonals.

6. Prepare a collection of puzzles that are suitable for use on a bulletin board as the "Puzzle of the Week."

7. Cut out some equilateral triangles and regular hexagons with sides of the same length. Then use them to form two different semiregular tessellations.

8. Draw the 12 different pentominoes on graph paper. Then show how each one of them can be used to form a tessellation.

READINGS AND REFERENCES

1. The 1980 Yearbook of the National Council of Teachers of Mathematics is titled *Problem Solving in School Mathematics*. Read and report on Chapter 17, "Problem Solving through Recreational Mathematics," by Kevin Gallagher.

2. Review the text *Teaching Mathematics in the Secondary School* by Alfred Posamentier and Jay Stempleman, Columbus, Ohio: Charles E. Merrill Publishing Co., 1986. In particular, prepare reports on three of the enrichment topics described in the second part of this text, "Enrichment Units for the Secondary School Classroom."

3. Review issues for the past year of the magazine titled *Games*, and report on any recreational items presented that are suitable for the mathematics classroom.

4. Read about the artist Maurits Escher. Copy several of his drawings that are based on tessellations illustrating the basic polygons from which the figures are developed.

5. An excellent source of mathematical recreations is any of the texts written by Martin Gardner and based on his columns from *Scientific American*. Prepare a report on mathematical games and puzzles as found in one of the following of his texts:

   Martin Gardner's Sixth Book of Mathematical Games from Scientific American, San Francisco: W. H. Freeman and Company, 1971.

Mathematical Carnival, New York: Alfred A. Knopf, 1975.

Mathematical Circus, New York: Alfred A. Knopf, 1979.

6. Read *How to Enrich Geometry Using String Designs*, a 1986 publication of the National Council of Teachers of Mathematics by Victoria Pohl. Construct at least one of the designs suggested by the author.

7. Read *Topics for Mathematics Clubs*, a 1983 publication of the National Council of Teachers of Mathematics by LeRoy Dalton and Henry D. Snyder. Select two of the chapters and prepare reports on the topics presented that are suitable for a secondary mathematics class or club.

8. Read the 1985 publication of the National Council of Teachers of Mathematics titled *Learning and Mathematics Games* by George Bright, John Harvey, and Margariete Wheeler. Report on their research concerning the use of games to promote learning in mathematics.

9. Review the 1984 publication of the National Council of Teachers of Mathematics titled *Student Merit Awards: High School*, edited by Leroy Sachs. Discuss the manner in which this publication can be used in secondary mathematics classes, and report on two of the suggested topics.

Classroom Aids and Activities: Numerical Concepts

Chapter 5

This chapter deals with aids and activities for arithmetic topics. Specific attention is given to fractions, decimals, and percents along with computational curiosities and number pattern experiments. Sample student worksheets are illustrated and simple classroom experiments described. Detailed construction steps appear as needed for the aids discussed in the chapter along with suggestions for their uses and possible extensions. Problem-solving experiences are emphasized throughout.

5.1   MOTIVATING NUMERICAL CONCEPTS

Once the various arithmetic skills have been introduced and developed for students, they need to be maintained throughout the rest of their mathematical training. This can often be a very difficult task since it requires a new look, a fresh approach, to familiar topics. Reviewing and drilling on the skills in the same way that they were introduced is to the student unimaginative, unmotivating, and all too often unsuccessful.

Use Visual Aids

Visual aids can frequently serve as the vehicle to motivate review, disguising an otherwise recognizable subject. The teacher who wants to spend a few minutes at the beginning of class reviewing whole-number computation could start with a number such as the date. Here is an example:

31487   (3/14/87 for March 14, 1987)

The first task is to get the class to recognize the number. Next the teacher brings out a set of cards numbered with these digits. The cards are mixed up and a student selects one. Then each student or team sees if it can be the first to use the other four cards with the fundamental operations to get the number selected. If the same card is selected a second time, a new answer must be given.

Here are some possible answers for the cards 3, 1, 4, 8, and 7, assuming that 7 is selected as the objective card.

$$8 + 3 - (1 \times 4) = 7$$
$$3(8 \div 4) + 1 = 7$$
$$3 \times 1 + 8 - 4 = 7$$
$$(8 + 1) \div 3 + 4 = 7$$

3 1 4 8 7

Now let each student use his or her own birthday for more practice. Some will have 4 digits instead of 5 and others will have 6. Some will contain zeros and some will repeat digits. Some sets will be easy to work with while others will be hard or impossible. It is a simple idea, a simple aid, a simple activity—but far more motivating than a drill sheet of problems.

Manipulate Geometric Models

Students need to see numerical equivalences emerge out of the manipulations of geometric models. Start with a square. Fold a pair of opposite vertices together and then fold them to the center point of the square. For a unit square with area 1, the first step gives $\frac{1}{2}$ and the second step gives $\frac{3}{8}$.

What fractional part results if the remaining two vertices of the square are now folded to the center point as well?

Pose Challenging Problems

Problems that reach beyond the routine often stimulate new interest in arithmetic computation.

A man was born in a year that was a perfect square. How old is he now?

Have your students guess first, and then use trial and error to find the last year that was a perfect square. Subtract from the current year to find the age.

For a more challenging adaptation of the same kind of problem, pose this question and discuss some of the possible answers.

> In recording a woman's year of birth and death, one was a perfect square and the other was a perfect cube. How old was she when she died?

A calculator can be a handy tool here. Show that the last two possible ages are 36 and 47.

5.2 FRACTIONS AND DECIMALS

Many students have difficulty understanding fractions and decimals and their related computational algorithms. Physical models and manipulative experiences often give these concepts and skills fuller meaning.

Recognizing Fractional Parts

The initial concept of a fraction is that of a fractional part of a whole, a geometric concept. If a student lacks this concept and yet is thrust into the algorithms of computation, success may be hard to come by. At all levels, take the time to relate fractions to geometry. This will allow for concrete experiences to bridge the gap to the abstractions inherent in the numerical symbols.

> Fold a square piece of paper along its diagonals. Each small triangle formed is what fractional part of the square? Since 4 small congruent triangles result, each must represent $\frac{1}{4}$ of the original square.

> The small square shown is what fractional part of the larger square? Ask first and then show that the answer must be $\frac{1}{2}$ by folding each corner of a large paper square to its center.

The small square is what fractional part of the larger square in this tangram puzzle with its seven pieces? If your eye can see a possible geometric subdivision of the figure into 16 small congruent triangles, then you know the small square is clearly $\frac{1}{8}$ of the larger.

Forming Fractions

Write four different single-digit numbers on cards. Use them for the numerators and denominators of two fractions.

How many different fractions can be formed with single-digit numerators and denominators using these cards?

Can two equivalent fractions be formed? If not, which two are nearest in value?

Which fraction has the greatest value? the least? Which two have the greatest sum? the least?

Students all too often only see fractions as single entities and not as ratios of numbers that can change. By moving the cards around, the students have a physical model to recall mentally later when asked what happens to a fraction when the numerator or denominator increases or decreases.

Adding Fractions

Fold a long rectangular strip into thirds vertically. Fold another the same size into sixths. Fold another into halves.

Tear off $\frac{1}{3}$ of the first strip and $\frac{1}{6}$ of the next; show that they equal $\frac{1}{2}$.

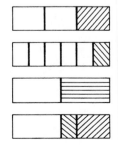

Use the same model to show these relationships:

$$\frac{1}{2} + \frac{1}{6} = \frac{2}{3}$$
$$\frac{1}{2} - \frac{1}{6} = \frac{1}{3}$$

Multiplying Fractions

Take a rectangular piece of paper. Fold it in thirds one way and then fold it in half the other way. Each of the six resulting sections is $\frac{1}{3} \times \frac{1}{4}$ and hence the product must be $\frac{1}{6}$.

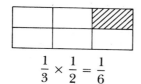

$$\frac{1}{3} \times \frac{1}{2} = \frac{1}{6}$$

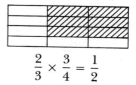

$$\frac{2}{3} \times \frac{3}{4} = \frac{1}{2}$$

Notice how a second similar folding can be used to show $\frac{2}{3} \times \frac{3}{4}$ geometrically.

Naming Decimals

Start with three cards on which you write the numbers 1, 2, and a decimal point. Name all the decimals that can be formed using the cards in any order.

For a more challenging version, add another card with a different digit, use any or all of the cards, and list all the decimals in order.

$$\boxed{1}\ \boxed{2}\boxed{.}\ \boxed{3}$$

The length of the list may surprise you.

.1	.12	.123	.13	.132	.2
.21	.213	.23	.231	.3	.31
.312	.32	.321	1	1.2	1.23
1.3	1.32	2	2.1	2.13	2.3
2.31	3	3.1	3.12	3.2	3.21
12	12.3	13	13.2	21	21.3
23	23.1	31	31.2	32	32.1
123	132	213	231	312	321

Ordering Decimals

Write a set of decimals on some strips of paper. Give students the decimals, one by one, and have them stand in the correct order, largest first, in front of the room. Choose the decimals at the appropriate level.

Upper level students will find a set such as the following both interesting and challenging.

This activity can serve as a change of pace, but even more it can dramatically illustrate and emphasize problems and difficulties in ordering decimals. The activity is readily adaptable to the review of fractions, percents, and integers.

Multiplying Decimals

Start with four digits on cards. Which digits should go where to make the indicated products the largest possible?

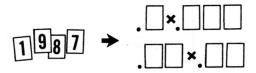

Moving the cards allows for visual reordering without repetitions, thus facilitating trial and error and analysis. For some, a calculator would be a useful tool. As a variation, do the same thing again but this time look for the smallest possible products.

Displaying Decimals

Keep a roll of machine tape and a marketing pen available when teaching decimals. These aids can be used to make some dramatic and powerful impressions about the nature of infinite decimals, both repeating and nonrepeating.

Write a great many digits on the paper tape. Then, at just the right time in class, unroll the tape across the room or tack it on the bulletin board.

Here, as examples, are two infinite and repeating decimals:

$$\frac{1}{7} = 0.142857142857142857142857142857142857142857142857\ldots$$

(repeats in blocks of 6 digits)

$$\frac{1}{17} = 0.05882352941176470588235294117647058823529411764 7\ldots$$

(repeats in blocks of 16 digits)

Ask Some Thought-Provoking Questions

What is the one hundredth digit in the decimal for $\frac{1}{7}$? for $\frac{1}{17}$?   (8, 8)

How many digits precede the one hundredth 7 in the decimal for $\frac{1}{7}$? for $\frac{1}{17}$?   (599, 799)

What is the last digit in the decimal for $\frac{1}{7}$? for $\frac{1}{17}$?

(There is none. No infinite decimal has a last digit.)

When discussing the nonrepeating nature of the decimal representation for the irrational number π, many students nod in agreement but do not really believe that it *never* repeats. The message is much more impressive and lasting when the students can actually see and study 100 or so of the digits for repeating patterns. All the teacher needs is a strip of paper and the time to copy the digits down.

π = 3.14159 26535 89793 23846 26433 83279 50288 41971
 69399 37510 58209 74944 59230 78164 06286 20899
 86280 34825 34211 70679 82148 08651 32823 06647
 . . .

Nonrepeating decimals with properties such as the following can generate some interesting questions when a long sequence is displayed.

0.79779777977779777779777777977777779777777779777777779 . . .

How many digits precede the one hundredth 9?

(5050 sevens plus 99 nines, for 5149 digits)

Are there more 7's or 9's in the decimal?

(Particularly unsettling is the fact that the infinite decimal has no more 7's than 9's!)

5.3 PERCENT

Many students have trouble with percents both in the junior high and senior high schools. The National Assessment of Educational Progress tests clearly bear out this point. Perhaps part of the problem arises from the methods in which percents are usually taught. Heavy emphasis on computational algorithmic procedures often leaves the student without a solid understanding of the concept and with little skill in mental manipulation of percents. Many students cannot visually estimate percents because they seldom see the subject in a geometric light. Concrete aids, activities, and manipulative experiences related to percent often clarify and reinforce the concept, which to many is extremely abstract.

Presentation Techniques

Cut out two circular pieces of paper of the same size but different in color. Use a protractor to divide each into 10 equal sectors. Cut along one radius to the center of each circle. Insert one circle inside the other so that it can rotate exposing anything from 0% to 100% of that color. Using both marked sides, students can count off by 10% units to get a visual reinforcement of various percents. Using the unmarked sides, students gain valuable experience in visual estimation.

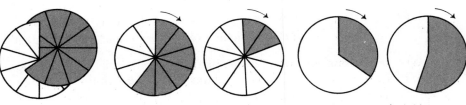

Insert one circle inside another	Rotate marked sides		Rotate unmarked sides	
	60% shaded	20% shaded	35% shaded	55% shaded

A number line can serve as a valuable model in doing percent problems mentally. For problems involving 25%, 50%, and 75%, begin by dividing the number line into four equal parts.

What is 75% of 12?

Place 0 under 0% and 12 under 100%. The number that would be under 75% is 9.

Note that the same picture can be used when mentally visualizing these problems as well.

9 is 75% of what number?
9 is what percent of 12?

This activity can be effectively presented using the overhead projector or the chalkboard. For an aid that students can handle, try graph paper or a ruler as a model.

Of course, the goal here is to encourage the students to think on the number line mentally. Time spent reviewing simple percent problems with models can significantly improve the students' understanding, comfort, and mental dexterity with percents later on.

Manipulative Experiences

Concrete activities in the hands of the students provide valuable experiences and allow you to present percent using a geometric rather than an arithmetic model. This frequently leads to interesting problem-solving situations.

Give each student a square piece of paper and have him or her label the vertices in order A, B, C, and D. Fold the square as indicated. What percent of the original square remains after each fold?

 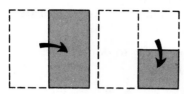

Fold A to B (50%) Fold B to C (25%)   Fold A to C (50%) Fold B to D (25%)

Now see how many students can give the correct percents of the original square remaining in these cases without actually folding the square. Use the folded square as a check.

Fold A to the midpoint of side AB.   (75%)
Fold A, B, C, and D to the center of the square.   (50%)
Fold A to C and then B to C.   (37-1/2%)

A percent chart drawn on a sheet of graph paper relates the topic of percent to proportions found in similar triangles. Have students label the chart as shown. Pivot a ruler or straightedge at the 0 point shown and move it along the base to serve as the tie line. The percentage can easily be read given the rate and base. Likewise, the rate can be read given the percentage and base.

What is 60 percent of 50?   What percent of 75 is 60?

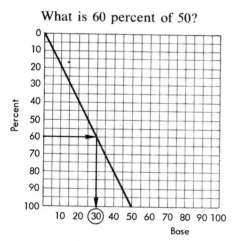   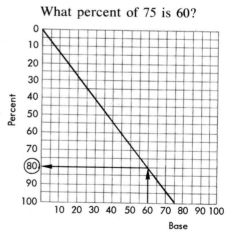

Locate the tie line through the base point of 50. Read across at 60 percent to find the percentage, 30.

Locate the tie line through the base point of 75. Read up at a percentage of 60 to find the percent, 80%.

Ask your students to describe how to find the base given the rate and percentage in an example such as this one:

40% of what number is 10?

Your students should recognize that the calculator is a valuable tool for checking percent computations and for exploring problem-solving situations without doing tedious calculations.

How long will it take for money to double itself at 8% compounded annually?

Here is a simple procedure that works on most inexpensive, nonscientific calculators.

Enter 1.08 and press the × key. Repeated pressing of the = key raises 1.08 to successive powers.

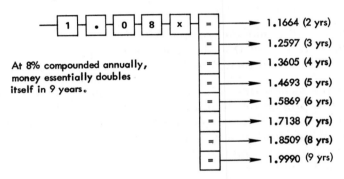

At 8% compounded annually, money essentially doubles itself in 9 years.

1.1664 (2 yrs)
1.2597 (3 yrs)
1.3605 (4 yrs)
1.4693 (5 yrs)
1.5869 (6 yrs)
1.7138 (7 yrs)
1.8509 (8 yrs)
1.9990 (9 yrs)

Use a similar process to show that the answer is 8 years 9 months at 8% compounded quarterly. Compounded daily using a 360 day year, the 8% rate doubles money in 8 years 240 days. (See page 150 for a further discussion of this topic using a microcomputer.)

5.4 COMPUTATIONAL CURIOSITIES AND GAMES

Many interesting computational curiosities are available to motivate instruction in mathematics. These range from items that the teacher can use for classroom demonstration to those that can be given to students in the form of a puzzle. The collection that follows is representative of the many topics in the area of computation that can serve to motivate students in the mathematics classroom while providing a review of arithmetic fundamentals.

1. **Digits in dates** One popular activity is to take the digits of a specific date and use these, together with operations, to generate various numbers. Usually the additional stipulation is stated that the digits must be used in the order in which they appear in the date under consideration.

As an example, consider the date 1776. Here are two ways to use the digits of this date to obtain the number 4.

$$17 - 7 - 6 = 4 \qquad (17 + 7) \div 6 = 4$$

An interesting project is to attempt to represent as many numbers as possible from 0 through 100 using these four digits in order. Depending upon the level of ability of the class, various operations may be permitted.

$$
\begin{array}{ll}
0: & 1^7 - 7 + 6 \\
1: & 1^{776}
\end{array}
\Bigg\} \quad \text{with exponents}
$$

$$
\begin{array}{ll}
2: & 1 + \sqrt{7 \times 7} - 6 \\
3: & \sqrt{1 + 7 + 7 - 6}
\end{array}
\Bigg\} \quad \text{with square roots}
$$

$$
\begin{array}{ll}
4: & |1 + (7 \div 7) - 6| \\
5: & |1 - 7| - 7 + 6
\end{array}
\Bigg\} \quad \text{with absolute values}
$$

$$
\begin{array}{ll}
6: & 1 \times (7 - 7)! \times 6 \\
7: & 1 \times 7 \times (7 - 6)!
\end{array}
\Bigg\} \quad \text{with factorials}
$$

$$
\begin{array}{ll}
8: & 1 + (7 \div 7) + 6 \\
9: & (1 + 7 + 7) - 6
\end{array}
\Bigg\} \quad \text{with fundamental operations only}
$$

This item can be used in a variety of ways. For example, it can be given as an assignment, it can be used as the basis for a race between chosen teams, it can form the basis for a bulletin board display, and so forth. It is most appropriate to use near the start of a new year, using the digits of the forthcoming year.

2. **Four fours** Closely related to the preceding item is one that requires as many numbers as possible to be represented using four 4's and any available operation. Here, for example, are representations for the numbers 0 through 10.

$$
\begin{array}{ll}
0: & 44 - 44 \\
1: & 44 \div 44 \\
2: & (4 \div 4) + (4 \div 4) \\
3: & (4 + 4 + 4) \div 4 \\
4: & 4 + 4(4 - 4) \\
5: & 4 + 4^{(4-4)} \\
6: & \sqrt{4 \times 4} + 4 - \sqrt{4} \\
7: & 44 \div 4 - 4 \\
8: & (4 \times 4) - 4 - 4 \\
9: & 4 + 4 + (4 \div 4) \\
10: & (44 - 4) \div 4
\end{array}
$$

3. **Coded messages** Students can be motivated to review arithmetic in disguise through the use of secret messages that they can uncover by doing the appropriate problems.

Work each problem. Then find the letter of the alphabet that corresponds to the answer. For example, the first problem is $126 \div 9 = 14$, and the fourteenth letter of the alphabet is N. After all problems are completed, the message is read down in the last column.

PROBLEM	NUMBER	LETTER
$126 \div 9$	14	N
75% of 20		
$\frac{2}{3} \times 12$		
$101 - 86$		
$\sqrt{169}$		
$5^2 \div 5$		
$(7 \times 5) - (4 \times 3)$		
$0.45 \div 0.03$		
2×3^2		
$(11 \times 11) \div 11$		

Depending upon the level of ability of the class, the problem can be made more or less difficult and can be designed to include extensive work with fractions, decimals, and percents.

4. **One through nine** A difficult but interesting project is to use all nine digits from 1 through 9 to represent 100. Challenge your class to find as many different ways as possible to do this, and make a collection of these on a bulletin board. Here are several examples.

$$1 + 2 + 3 + 4 + 5 + 6 + 7 + (8 \times 9) = 100$$
$$123 - 45 - 67 + 89 = 100$$
$$1 + 2 + 3 - 4 + 5 + 6 + 78 + 9 = 100$$
$$(56 + 34 + 8 + 2) \times (9 - 7 - 1) = 100$$
$$62 + 38 + [(1 + 7) \times (9 - 5 - 4)] = 100$$

To make it even more interesting the teacher may wish to restrict the use of the nine digits to their natural order, as in the first three examples above.

5. **Card games** Mark ten 3×5 file cards with the numbers 1 through 10. Shuffle them. Deal out five cards and use each of those values once, with any of the four fundamental operations, to give the value on the next card turned over. Play the game with the class or let students play against each other. Work for speed and accuracy. The first student to find a combination for the answer wins, or every student with a correct solution within a given time wins a point.

$$(8 - 6 - 1)(3 + 4) = 7$$
$$(8 - 1)(3 + 4 - 6) = 7$$
$$1(8 + 34)/6 = 7$$

This game can give some good mental practice in problem solving by the trial-and-error method. By having the students express the process algebraically, additional practice can be gained in the proper use of grouping symbols. This is an excellent game to help develop better number awareness.

5.5 NUMBER-PATTERN EXPERIMENTS

One of the objectives in teaching mathematics is to develop number awareness. Experiments that offer physical situations leading to the recognition of number patterns and sequences meet this need. They involve the individual student with a hands-on activity, a direct instruction leading to a countable result, a chance to discover a pattern and offer educated guesses at successive results before actual verification, and a challenging opportunity to generalize algebraically where appropriate. These are valuable ingredients in a meaningful and motivating classroom experiment.

Some of the best classroom experiments involve student manipulatives made of the simplest materials based on straightforward directions and clear objectives. Such activities are all the better when they capture the imagination and contain unexpected surprises that lead to new observations, pose new questions, and force new strategies.

Consider the following classroom experiment that involves just a sheet of paper in each student's hands. By repeatedly folding the paper in half, some interesting number-pattern questions emerge.

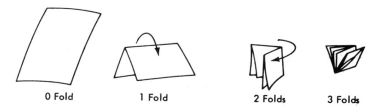

| 0 Fold | 1 Fold | 2 Folds | 3 Folds |

What is the relationship between the number of folds and the thickness?

Have the students fold and count. Encourage careful and systematic tabulation. Let the students make all the discoveries.

Number of folds	0	1	2	3	4 ...
Number of sheets thick	1	2	4	8	16 ...

Undoubtedly, many students will be able to visualize the results without actually folding the paper. The thickness doubles with each successive fold. Thus with n folds, the thickness is 2^n. Have the students compute the number of folds needed to get the number of sheets in a 500-page book. Recall that a 500-page book uses only 250 sheets. Give the paper to some students and see if they can fold it to this thickness. They will probably get no more than 7 folds. Invariably, some student will suggest using a larger sheet of paper to get more folds. Be ready with some large pages from a newspaper. Let the students try. The results will undoubtedly surprise both them and the rest of the class.

The examples of experiments in number patterns that follow include some with detailed descriptions and analyses and brief suggestions for others. They are probably best used one or two at a time throughout the year rather than all at one time. Some are more suited for lower grades and slower classes; others can be challenging for the better secondary students. But in all cases, the student should take educated guesses and try to verify them with the material supplied. Students should be encouraged to discover and verbalize the pattern illustrated. However, many of the generalizations of the nth terms can be left for the better students.

EXPERIMENT 1   Folding Paper

Material

One sheet of paper per student.

Directions

1. Fold the paper once, open it up, and record the number of regions.
2. Fold again for the maximum number of regions possible.
3. Repeat the process again for 3 folds. Remember, open the paper flat before each new fold and always fold for the *maximum* number of regions possible.
4. Try to discover the number sequence and predict the result for 4 folds. Check your answer by folding again and counting regions.
5. Can you generalize the sequence for n folds?

Analysis

The maximum number of regions is always formed if the new fold cuts every existing fold without passing through any intersection or endpoint.

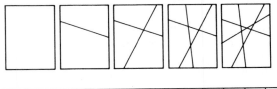

Number of folds	0	1	2	3	4
Maximum number of regions	1	2	4	7	11

1   2   4   7   11   16
   1   2   3   4   5

The first differences form the counting numbers. Hence for 5 folds, the answer is the sum of the first five counting numbers plus 1.

$$(1 + 2 + 3 + 4 + 5) + 1 = 16$$

For n folds the maximum number of regions is the sum of the first n counting numbers plus 1.

$$\frac{n(n + 1)}{2} + 1$$

For a simpler initial experiment, consider the same problem but always fold for the *minimum* number of regions possible. Try counting the number of polygons formed at each stage as well.

Here the minimum number of regions is always 1 more than the number of folds. As a generalization, for n folds, the minimum number of regions is $n + 1$.

The corresponding number of polygons formed, using both folds and edges for sides, generates this important sequence for 0, 1, 2, 3, 4, . . . folds.

$$1, 3, 6, 10, 15, \ldots$$

This should be readily recognized as the sums of successive counting numbers. For n folds, the number of polygons formed is $n(n + 1)/2$.

EXPERIMENT 2   Cutting String

Material

Pieces of string and scissors.

Directions

1. If you know the number of cuts in a string, do you know the number of pieces? Try completing the table without cutting the string first.

number of cuts	0	1	2	3	4	5
number of pieces						

If there are n cuts in the string, how many pieces will there be?
If there are n pieces, how many cuts were made?

2. Fold the string as shown before you cut. Do you know the number of pieces that will be formed this way from a given number of cuts? Again, try to complete the following table without cutting the string.

number of cuts	0	1	2	3	4	5
number of pieces						

If there are n cuts, how many pieces will there be?

If there are n pieces, how many cuts were made?

3. Loop the string through the scissors as shown. Find how many pieces you get for each number of loops.

 0 Loop   1 Loop

number of loops	0	1	2	3	4	5
number of pieces						

If there are n loops, how many pieces will there be?

If there are n pieces, how many loops were cut?

Analysis

1. Each cut produces an extra new piece. Hence for n cuts there will be $n + 1$ pieces. To get n pieces, $n - 1$ cuts are needed.
2. Here each cut produces 2 new pieces. Hence for n cuts there will be $2n + 1$ pieces. Note that the number of pieces must always be odd. To get n pieces, with n odd, make $(n - 1)/2$ cuts.
3. In this case we are counting loops and pieces rather than cuts and pieces. Starting with 0 loops is the same as starting with 1 cut which gives 2 pieces. One loop is the same as 2 cuts and gives 3 pieces. Each additional loop means an additional piece when cut. For n loops, there are $n + 2$ pieces. To get n pieces, cut through $n - 2$ loops, but remember that n cannot be less than 2.

EXPERIMENT 3 Counting Regions

Material

A piece of string.

Directions

Place the string on the desk so that it crosses itself just once. How many regions are formed? Move the string so that it crosses itself twice. Now how many regions are formed? Is there any other way that you can lay the string so that it crosses itself twice but forms a different number of regions? Try the same for three and four crossings. Can you discover a pattern?

Analysis

The number of bounded regions formed is the same as the number of crossing points. But there is always an additional exterior unbounded region. Thus, for n crossing points, there are $n + 1$ regions.

EXPERIMENT 4   Circles

Material

Compasses, rulers, and paper.

Directions

Draw some circles on your paper and use them to try to discover the number patterns in each of these situations.

Find the maximum number of regions possible in a circle for various numbers of radii, diameters, chords, and tangents.

Radii	Diameters	Chords	Tangents

Radii	Diameters	Chords	Tangents
1	1	1	1
2	2	2	2
3	3	3	3
4	4	4	4
5	5	5	5

Analysis

1. 1, 2, 3, 4, 5

   With 1 radius there is 1 region. Each additional radius, wherever it is drawn, increases the number of regions by 1. For n radii there must be n regions.

2. 2, 4, 6, 8, 10

   With 1 diameter there are 2 regions. Each additional diameter, wherever it is drawn, increases the number of regions by 2. For n diameters there must be $2n$ regions. It follows that the number of regions will always be even.

3. 2, 4, 7, 11, 16

   To produce the maximum number of regions, care must be taken as to where the successive chords are drawn. Each new chord must intersect all other chords drawn but not at any existing intersection or endpoint. For n chords, the maximum number of regions possible is $n(n + 1)/2 + 1$.

4. 1, 1, 1, 1, 1

   Since tangents to a circle contain points of the circle but never pass through its interior, the number of regions inside the circle remains 1 for any number of tangents. However, an interesting counting pattern emerges when all regions inside and outside the circle are counted. In this case these would be the tabled values:

   3, 6, 10, 15, 21

   For n tangents to a circle the maximum number of regions possible in the plane is $(n + 1)(n + 2)/2$, which is the sum of the first $n + 1$ counting numbers.

Another challenging version of this last experiment counts the maximum number of regions possible in the plane using increasing numbers of secants. The results are 4, 8, 13, 19 for 1, 2, 3, 4 secants. Successive second differences increase by 1 each time. For n secants, a maximum of $(n + 2)(n + 3)/2 - 2$ regions are possible in the plane.

EXPERIMENT 5 Figurate Numbers

Materials
Graph paper to facilitate drawing the similar figures.

Directions
1. Copy the figures shown. Then sketch the next one.
2. Count the dots in each figure.
3. Find the first differences. What pattern do you get?
4. Find the second differences. What pattern do you get?
5. What is the nth figurate number in the set?

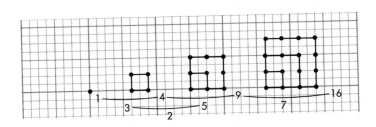

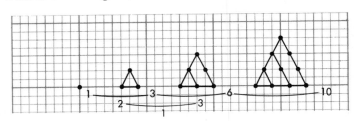

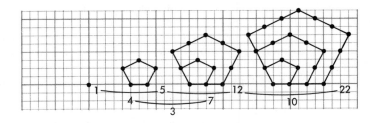

Triangular numbers	1,	3,	6,	10,	15,	21,	28,	36, . . .
first differences		2	3	4	5	6	7	8 . . .
second differences			1	1	1	1	1	1 . . .

The nth triangular number is $n(n + 1)/2$.

Square numbers	1,	4,	9,	16,	25,	36,	49,	64, . . .
first differences		3	5	7	9	11	13	15 . . .
second differences			2	2	2	2	2	2 . . .

The nth square number is n^2.

Pentagonal numbers	1,	5,	12,	22,	35,	51,	70,	92, . . .
first differences		4	7	10	13	16	19	22 . . .
second differences			3	3	3	3	3	3 . . .

The nth pentagonal number is $n(3n - 1)/2$.

Note that the second differences are especially interesting. For a given set, they are constant and take on the value of 2 less than the number of sides in the figurate number.

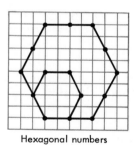

Hexagonal numbers

Students might be interested in also exploring hexagonal figurate numbers. The first three are shown in the figure as 1, 6, 15. Others can be computed quickly using the fact that all second differences are 4. (See page 148 for a further discussion of figurate numbers using a microcomputer.)

The following experiments focus on discovering number patterns using geometric figures and manipulatives. They are valuable geometric as well as numerical experiences.

EXPERIMENT 6   Counting Squares

Material
Some square pieces of paper.

Directions
Repeatedly divide a square into smaller and smaller squares. Count the number of the smallest squares formed after each successive division.

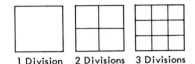

1 Division   2 Divisions   3 Divisions

Repeat the process only this time count all squares of all sizes.

Analysis
The pattern for the number of small squares is readily apparent. For n subdivisions per side the number is n^2. This number pattern gives a clue to the more challenging and beautiful pattern for the total number of squares. The first differences are the successive squares. Hence the number of squares of all sizes for n subdivisions is the sum of the squares of the first n counting numbers.

number of divisions	1	2	3	4	5
number of small squares	1	4	9	16	25
number of all squares	1	5	14	30	55

What is the total number of squares of all sizes on a checkerboard?

EXPERIMENT 7   Counting Triangles

Material
A triangular piece of paper.

Directions
Repeatedly fold a triangle through one of its vertices. Count the total number of triangles formed after each fold. Try to discover the pattern.

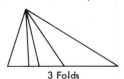

Analysis
3 Folds

number of folds	0	1	2	3	4	5
number of triangles	1	3	6	10	15	21

It is surprising how many different places the set of the sums of successive counting numbers appears.

5.6 MAGIC SQUARES ACTIVITIES

Many interesting activities involving arithmetic review can be presented within the format of magic squares. A magic square is a square array of numbers where the sums of the entries in each row, each column, and each diagonal are all the same.

This 3×3 magic square has a magic constant of 201.

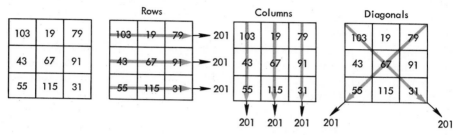

Figure 5.3

Every magic square has eight different rotations and reflections. Can you find the other seven positions for the nine numbers in the magic square above?

New magic squares can be formed from existing ones using one or more of the four fundamental operations. Notice that the simple 3×3 magic square using the numbers 1 through 9 can be transformed into the one shown above by changing every entry x to $12x + 7$.

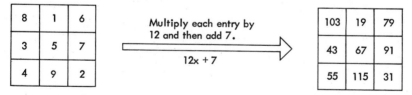

What algebraic transformation will change the above magic square on the right, back to the one on the left?

Magic squares can involve decimals, fractions, integers, and variables as well as whole numbers. Verify that each of these is a magic square.

1.75	0.5	4	2.25
3	3.25	0.75	1.5
0.25	2	2.5	3.75
3.5	2.75	1.25	1

2/3	1/12	1/2
1/4	5/12	7/12
1/3	3/4	1/6

-1	-8	15	8	1
-7	11	9	2	0
12	10	3	-4	-6
6	4	-3	-5	13
5	-2	-9	14	7

N+2	N+3	N-2
N-3	N+1	N+5
N+4	N-1	N

Here is a method that can be used to construct odd-ordered magic squares, those with an odd number of rows and columns.

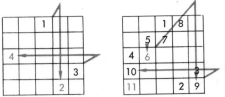

17	24	1	8	15
23	5	7	14	16
4	6	13	20	22
10	12	19	21	3
11	18	25	2	9

Step 1: Write the first number, 1, in the middle of the top row. Fill in successive counting numbers moving diagonally upward and to the right. When the next position is off the square, enter the number at the opposite end of the next row or column. Note how this has been done for the entries 2 and 4.

Step 2: Continue working diagonally upward and to the right. When you come to a position already occupied, enter the number in the space immediately below the last one filled. Note how this has been followed for the numbers 6 and 11.

Step 3: The space that follows the upper right-hand corner is the lower left-hand corner. Since this space is occupied by 11, the 16 is placed below 15 following the rule from step 2.

Step 4: Continue the process until the 5 x 5 magic square is finished. A 5 x 5 magic square has 5 x 5, or 25, entries. Since 1 was the first number, 25 is the last.

Check to see that the completed array is indeed a magic square.

Using this method, construct a 7 x 7 magic square starting with a 1 in the middle of the first row. How would you construct a 7 x 7 magic square starting with a 100 in the middle of the first column?

The following five pages illustrate how individual Student Worksheets can be prepared for exploring properties of magic squares. The suggestions can be readily modified and extended to fit other needs. Worksheet 1 deals with rotations and reflections of a given magic square and involves visualization skills. Worksheet 2 calls for the completing of magic squares using problem-solving strategies. Worksheet 3 reviews computational skills and leads to a generalization concerning transforming one magic square into another. Worksheet 4 ties the topic to algebraic symbolism and proof. Worksheet 5 investigates properties of a magic cube.

Rearranging entries in a magic square

Many different numbers can be used in forming a 3 x 3 magic square. Likewise, many arrangements can be made from the same set of numbers in a magic square.

This simple 3 x 3 magic square uses the numbers 1 through 9. The numbers in each row, column, and diagonal add to 15.

See if you can complete these magic squares using the same set of numbers, 1 through 9. Remember, the numbers in each row, column, and diagonal still must add to 15.

 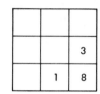

Cut out a square piece of paper and mark it with the original magic square. Turn the paper over and mark it on the back so that each number is in exactly the same square as in the front. Now see if you can get the magic squares above by simply rotating and flipping over this cutout. For example, if you flip over the original magic square about its vertical axis, you get the square in the upper left-hand corner. Find the one remaining arrangement possible with the numbers 1 through 9 that still forms a magic square.

Completing magic squares

Study these incomplete square arrays. Then try your problem-solving skills in completing them so that they become magic squares.

16	2	12
	18	

8		14	3
15	2	9	4
	11		13

13	12	6	25	
	14	8	7	
		15		3
4			16	
11	5	24		17

		53	10
13		4	7
	9	11	
51	12		

1		
3/8		
1/2		1/4

1 1/6		
	11/12	
1		2/3

An odd-ordered magic square has an odd number of rows and columns. Each one also has a special relationship between its magic sum and its center value. See if you can discover this property and use it to complete these magic squares.

32		24
	20	

		7
		21
		11

12		4
		8

Operating on magic squares

What happens when you add to or multiply each entry in a magic square with a constant? Will you get another magic square?

Start with the magic square shown. In each case add the constant given to each entry to form a new array. Then check to see if it, too, is a magic square.

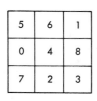

	Add 25	Add 1 1/4	Add 1.5

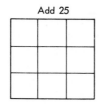 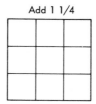 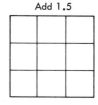

5	6	1
0	4	8
7	2	3

Does it appear that when you add the same number to each entry in a magic square, another magic square is formed?

Now multiply each entry in this magic square by the numbers given.

20	0	14
5	11	17
8	23	2

Multiply by 9	Multiply by 3/4	Multiply by 1.3

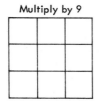 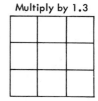

Does it appear that when you multiply each entry in a magic square by the same number, another magic square is formed?

The entries in this magic square are integers. Show that the numbers in each row, column, and diagonal add to 0. Then perform the operations indicated. In each case, see if the resulting array is also a magic square.

-3	2	1
4	0	-4
-1	-2	3

Subtract -2	Multiply by -3	Add -3/4 then multiply by -1/2

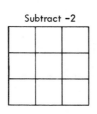 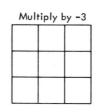

Choosing your own magic sum

Here is a very special 4 × 4 magic square that you can build with any magic sum. First select any number for the magic sum. When you complete the square below using this value for S, all ten rows, columns, and diagonals will have this sum S.

Let S = _____. Then compute the values given and enter them in the positions shown. (For all cells to be positive, S must be 22 or more.)

$S - 18 =$ ——— $\longrightarrow$

$S - 19 =$ ——— $\longrightarrow$

$S - 20 =$ ——— $\longrightarrow$

$S - 21 =$ ——— $\longrightarrow$

5	10	3	
4		6	9
	1	12	7
11	8		2

Check your work with these values. Add along each row, each column, and each diagonal. Each sum should equal the magic sum S that you chose.

Here is why this will always work. Enter the algebraic expressions for the missing values using the variable S.

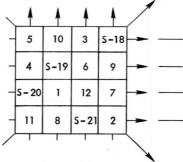

5	10	3	$S-18$
4	$S-19$	6	9
$S-20$	1	12	7
11	8	$S-21$	2

Find the sum for each row, each column, and each diagonal and write it in the space provided. What is each sum?

Try the same process for this 5 × 5 magic square.

Let S = _____. Then compute the values given and enter them in the positions shown. (For all cells to be positive, S must be 45 or more.)

$S - 40 =$ ——— $\longrightarrow$

$S - 41 =$ ——— $\longrightarrow$

$S - 42 =$ ——— $\longrightarrow$

$S - 43 =$ ——— $\longrightarrow$

$S - 44 =$ ——— $\longrightarrow$

11	18		2	9
17		1	8	15
	5	7	14	16
4	6	13	20	
10	12	19		3

Show algebraically why this method must yield a magic square with the magic sum S.

What is magic about this cube?

Here is a pattern for a model of a 3 x 3 x 3 magic cube. When assembled, it appears to be formed from 27 smaller cubes, each bearing a single number from 1 through 27.

			10	24	8						
			23	7	12						
			9	11	22						
10	23	9	9	11	22	22	12	8	8	24	10
26	3	13	13	27	2	2	25	15	15	1	26
6	16	20	20	4	18	18	5	19	19	17	6
			20	4	18						
			16	21	5						
			6	17	19						

1. Cut out and assemble the cube.
2. Find the face with 1 on it. Write the three numbers in each of the two rows that contain the 1. Add each set of numbers. Are both sums the same? Now add the three numbers in each of the four other rows on the same face. Do you always get the same sum?
3. Find the face with 27 on it. Add each of the six rows of three numbers on this face. What do you find?
4. There are 24 different rows of three numbers on the six faces of the cube. Can you find them all? Are all their sums the same?
5. One of the numbers from 1 through 27 is not marked on any face. What number is it? Where is the cube located with this number on it?
6. The missing cube is numbered 14 and it is in the very middle of the 3 x 3 x 3 cube. There are 13 different rows and diagonals in the cube that contain this small middle cube numbered 14. Can you find them all? Are all their sums the same?
7. Why do you think this 3 x 3 x 3 cube is called a magic cube?

Napier's Rods

Invented by the Scottish nobleman John Napier (1550–1617), these simple computing devices were in common use during much of the 1600s. They were designed to simplify the burdensome task of multiplication by the very man who later invented logarithms—which, in effect, translated multiplication problems into addition problems. The original rods were made from strips of wood or bone and small enough to carry in the pocket. Each rod had four sides with a scale on each side. By placing the appropriate rods side by side, you had a convenient computing device for multiplying quickly.

Napier's Rods can be an interesting and exciting topic for students at most levels of ability, especially when placed in their proper historical perspective. Several methods of constructing demonstration models are given here. However, this activity also offers an excellent opportunity for individual student involvement where each constructs and manipulates his or her own personal set of Napier's Rods.

Index	1	2	3	4	5	6	7	8	9	0
1	0/1	0/2	0/3	0/4	0/5	0/6	0/7	0/8	0/9	0/0
2	0/2	0/4	0/6	0/8	1/0	1/2	1/4	1/6	1/8	0/0
3	0/3	0/6	0/9	1/2	1/5	1/8	2/1	2/4	2/7	0/0
4	0/4	0/8	1/2	1/6	2/0	2/4	2/8	3/2	3/6	0/0
5	0/5	1/0	1/5	2/0	2/5	3/0	3/5	4/0	4/5	0/0
6	0/6	1/2	1/8	2/4	3/0	3/6	4/2	4/8	5/4	0/0
7	0/7	1/4	2/1	2/8	3/5	4/2	4/9	5/6	6/3	0/0
8	0/8	1/6	2/4	3/2	4/0	4/8	5/6	6/4	7/2	0/0
9	0/9	1/8	2/7	3/6	4/5	5/4	6/3	7/2	8/1	0/0

Construction

For the chalkboard, mark a large sheet of poster paper as shown, using a special color for the index numbers. Cut it into strips and set them on the chalk ledge or attach them to the board for demonstration.

For the overhead projector, copy the strips on a sheet of acetate, cut them out, and project them on the screen or board.

For individual use, supply students with copies as shown. Once they cut off the strips, they can then proceed at their own pace in exploring and using the activity. Loose strips can be kept in an envelope for future use.

Uses

1. Place the index rod alongside one of the others and show how the corresponding multiples are given. For example, with the 8-rod and the index, ask for such products as 7 × 8.

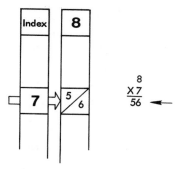

2. Repeat using two- and three-digit factors. For example, show 7 × 86 and 7 × 862. Explain how "carrying" works when adding along the diagonals. Let students read other products involving 8, 86, and 862 from the same settings.

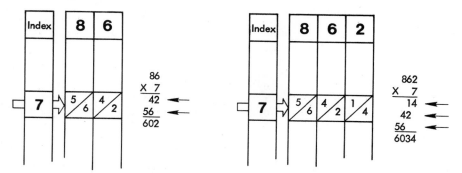

3. Allow students to suggest other multiplication problems, still with one-digit multipliers, and let others illustrate them at the board. See that some include the 0-rod.

4. Relate the results shown on the rods to the common computing algorithm now used for multiplying.

5. Let students discover how the rods can be used with two- and three-digit multipliers.

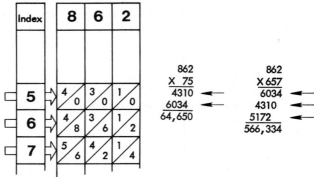

$$\begin{array}{r} 862 \\ \times\ 75 \\ \hline 4310 \quad\leftarrow \\ 6034 \quad\leftarrow \\ \hline 64{,}650 \end{array}$$

$$\begin{array}{r} 862 \\ \times 657 \\ \hline 6034 \quad\leftarrow \\ 4310 \quad\leftarrow \\ 5172 \quad\leftarrow \\ \hline 566{,}334 \end{array}$$

Suggested extensions

1. Students can make their own set using strips of ½-inch graph paper. Let them review the multiplication facts by filling in all the entries themselves and also by checking answers read from the rods.

2. Have students explain why there is no need for a 0 on the index strip. How do the partial products for 47 × 9532 compare with those for 407 × 9532 and 4007 × 9532?

Binary Cards

Here is an interesting classroom experiment that can be used to reinforce the concept of the binary system used in computers.

Materials   Index cards, paper punch, scissors.

Directions   As a first step, write the numbers from 0 through 15 in binary notation. Prefix zeros so as to have a four-digit numeral for each number.

BASE 10	BASE 2	BASE 10	BASE 2
0	0000	8	1000
1	0001	9	1001
2	0010	10	1010
3	0011	11	1011
4	0100	12	1100
5	0101	13	1101
6	0110	14	1110
7	0111	15	1111

Next have each student prepare a set of 16 index cards, with four holes punched in each and with one of the corners cut off for purposes of identifying the face of the card. Now the numbers from 0 through 15 are represented as in the following figures. In each case a hole alone represents 0, while a slot cut above the hole to the edge of the card represents 1. The number that each card represents should be written on the face of the card, as shown.

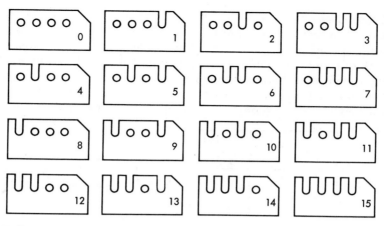

Follow the construction with these activities:

1. Place the cards in a stack and shuffle them well. How can all even-numbered cards be located? The even-numbered cards have 0's in the units place, the first hole on the right. Therefore, use a pencil or paper clip to push through this first hole. The cards lifted out will be the even-numbered ones. How can you sort out all the multiples of 4?

2. With only four sorts, any number from 0 through 15 can be located. Explain how to locate card 13 in just four sorts.

3. With only four sorts, put the shuffled cards back into numerical order from 0 through 15.

Analysis

1. Sort to remove those cards with 0's in the first two places from the right. These will be 0, 4, 8, and 12. How can the number 0 be sorted out?

2. Working from right to left,
   discard those pulled in the units place,
   keep only those pulled in the twos place,
   discard those pulled in the fours place,
   discard those pulled in the eights place.

3. Start with the first hole from the right. Place all the cards that lift up in front of the other cards. Then repeat this procedure for each of the three remaining positions in order. When finished, the cards will appear in numerical order.

Suggested extensions

Have students write instructions for locating a particular card. Let other members in class follow the instructions to find the card in question.

Ask how many cards and holes are needed to sort out 32 pieces of data. How many cards can be sorted using 10 holes?

Flow Charts

Parentheses in an arithmetic or algebraic expression help determine the order or sequence of operations to follow. Their importance is emphasized in this activity with the aid of flow charts.

Construction

Cut out a series of rectangular and circular pieces of paper and label them with operation commands and inputs. The specific samples illustrated make use of these commands and inputs. But they can be easily modified and expanded to include other inputs, such as decimals, fractions, and integers, and other operations, such as taking square roots. Selected pieces are attached to the bulletin board or chalkboard to form various flow charts.

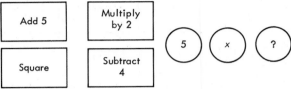

This activity is very effective when the pieces are constructed of acetate and arranged and rearranged on the overhead projector.

Uses

1. Students evaluate the output of each flow chart and express the sequence of operations using parentheses as needed.

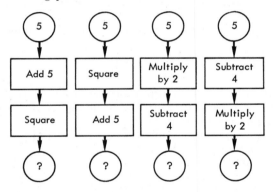

   By reordering the operations, expressions are formed with different values, thereby stressing the importance of the use of parentheses.

2. Arithmetic expressions are written on the board. Students then construct the corresponding flow charts from the various inputs and operation commands available. Here are some examples using pieces from the above set.

$$2 \cdot 5^2 \quad \text{and} \quad (2 \cdot 5)^2$$
$$2 \cdot 5 + 5 \quad \text{and} \quad 2(5 + 5)$$
$$5^2 - 4 \quad \text{and} \quad (5 - 4)^2$$

   Valuable experience in the use of parentheses and reinforcement of computational skills come from this simple activity.

3. Introducing a variable into the flow chart sets the stage for developing algebraic skills with parentheses. Repeat the methods suggested in steps 1 and 2, but use variables this time.

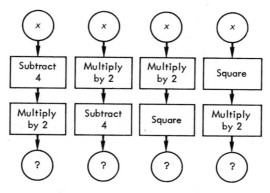

4. Step 3 stresses the formulation and meaning of algebraic expressions. Another different but very important skill is evaluating algebraic expressions. Use some of the flow charts constructed above and have students evaluate them for various values of the variable.

5. Flow charts with two or more operations are multistep problems. Have students make up some multistep word problems and construct the corresponding flow charts to emphasize the proper order of operations.

Suggested extensions

Place these operations and input on the board. Ask the students to arrange the pieces in as many different flow charts as possible, writing the correct arithmetic expression for each one. There are six possible orderings of the operation pieces; hence six resulting expressions can be formed and evaluated.

| Multiply by 2 | Subtract 4 | Square | 5 | ? |

$$(2 \cdot 5 - 4)^2 = ? \qquad 2(5 - 4)^2 = ? \qquad (2 \cdot 5)^2 - 4 = ?$$
$$[2(5 - 4)]^2 = ? \qquad 2(5^2 - 4) = ? \qquad 2 \cdot 5^2 - 4 = ?$$

Write the 24 expressions possible when a fourth operation such as "Add 5" is added to those given. Find out which expressions are equivalent.

Nomographs

The word *nomograph*, which comes from the Greek and means a written law, applies in mathematics to a graphic technique for computing and for the solution of specific equations. Nomographs can serve as the source of many experiments that lead to discoveries and reviewing skills. Students gain valuable experience in constructing these graphs and in reading their scales as well as in verifying the results of computational problems.

The simplest example is the parallel scales nomograph. Better students can construct one from scratch using a ruler. With slower students you may prefer that it be constructed on graph paper.

Each of these examples of nomographs involves addition in the form

$a + b = c$. The scales are parallel with equal spaces between them. Outer scales are calibrated alike while the middle scale is reduced to half-size.

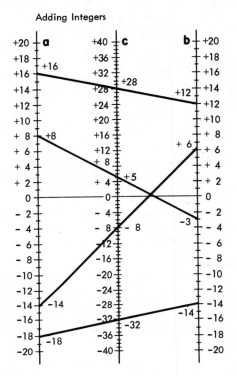

Adding Integers

This nomograph may be used to

1. find the sum of two integers
2. illustrate that each integer and its opposite add to 0
3. illustrate the commutative property for addition
4. relate subtraction of integers to addition

The theory behind the construction of parallel scales nomographs can be explored by your better students.

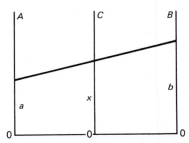

The segment x is the median of the trapezoid with parallel bases a and b. Its length is half that of the sum of the lengths of the bases. By calibrating the C scale to half that of the outer scales A and B, the C scale can be used to read the sum directly.

Not all nomographs have parallel scales. This one can be used to find k given a and b in the relationship $\dfrac{1}{a} + \dfrac{1}{b} = \dfrac{1}{k}$. It is often called a "spider" nomograph.

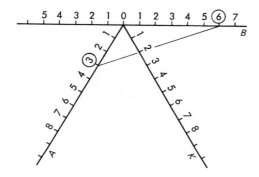

Here are three application problems that can be solved with this nomograph. Each solution can be read directly from the single tie line shown.

1. Find the sum of the reciprocals of 3 and 6. $\left(\dfrac{1}{3} + \dfrac{1}{6} = \dfrac{1}{2}\right)$

2. Pipe A fills a tank in 3 hours while pipe B fills the same tank in 6 hours. How long will it take for both pipes to fill the tank together?   (2 hours)

3. A 6-ohm and a 3-ohm resistor are wired in parallel. What is the single equivalent resistance?   (2 ohms)

5.8 OVERHEAD PROJECTOR IDEAS

The overhead projector can be an extremely effective teaching tool, especially when the transparencies are designed to capture the attention of the students. Make use of overlays, masks, color, and motion wherever possible. Several ideas are given here that can be used to present familiar computational review in a new format and with some added problem-solving twists.

Flow Charts with Changing Inputs, Operations, and Outputs

The flow chart format is readily adaptable to addition, subtraction, multiplication, and division using whole numbers, integers, and rational numbers expressed as fractions, decimals, or percents. A wide variety of problems can be generated easily and rapidly by sliding acetate strips across openings in a mask.

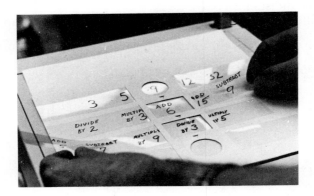

Construction

Step 1. Start with a piece of tagboard cut from a side of a file folder. Cut out two input/output circles and two instruction boxes each 1 inch high and 1½ inches long. Use a sharp knife or razor blade.

Step 2. Cut four strips of clear acetate 1¼ inches wide. Then cut a slot on each side of each opening as shown. Make the slots just wide enough to allow the strips to slide in them easily.

Step 3. Slide the strips through the slots so that they are on top on both sides and under where the openings have been cut.

Step 4. Label the input and operation strips for the use planned. Several suggestions are given.

Whole number review:

0	1	4	12	36
Multiply by 2	Add 6	Subtract 3	Divide by 2	Square
Add 8	Subtract 9	Divide by 3	Square	Multiply by 3

Integer review:

−1	+3	−6	+10	−15
Multiply by −2	Add −3	Subtract +6	Divide by +3	Square
Add +5	Square	Divide by −3	Subtract −2	Multiply by −4

Fraction review:

0	$\frac{1}{2}$	$\frac{3}{4}$	$\frac{1}{3}$	$1\frac{1}{8}$
Add $\frac{1}{2}$	Subtract $\frac{1}{8}$	Add $\frac{3}{4}$	Multiply by 2	Divide by 2
Multiply by $\frac{1}{3}$	Divide by $\frac{1}{2}$	Multiply by 3	Add $1\frac{1}{3}$	Subtract $\frac{1}{4}$

Uses

1. Vary the input values for given operations. Vary the operations for given input values. Work for speed and accuracy. Strive for mental computation.
2. For slower students, choose simpler numbers. Select fractions with the same denominator or simpler multiples.
3. Remove the input strip and insert a strip with selected output values. Here students gain the valuable problem-solving experience of working backwards using inverse operations.
4. At a more challenging level, include both input and output values and remove an operation strip. Have students discover possible operations that could be used.

Powers with Changing Exponents and Bases

Motion and change can be effectively illustrated using acetate dials mounted on a mask with common sewing snaps. The method illustrated here uses exponents but is adaptable to other applications as well.

Construction

Step 1. Cut a square out of a side of a file folder, and cut two acetate circles as shown.

Step 2. Mark the centers of the circles and their locations on the tagboard. Cut out small triangular holes at these points and then attach the dials using small sewing snaps.

Step 3. Mark selected values for the base and exponent on the acetate dials.

Uses

1. For a given base, vary the exponent. For a given exponent, vary the base. Again, stress speed, accuracy, and mental computation by the students.
2. Pose some problem-solving situations. Have students look for special properties and patterns in the powers. An interesting pattern that can be investigated here is that of the units digit in successive powers of a given base. Sequence the examples in this order.

Base 5   Units digits in successive powers: 5,5,5,5,5,5,5,5, . . .
       4   4,6,4,6,4,6,4,6, . . .
       3   3,9,7,1,3,9,7,1, . . .

Test understanding with extensions and generalizations.

What are the units digits for 5^{25}, 4^{25}, and 3^{25}?
What about 1, 2, 6, 7, 8, and 9 raised to the 25th power?

Modifications

At more advanced levels, make another dial using decimals and fractions for the base. Exponents can also be varied to include negative values.

Varying Word Problem Format

In this illustration, the dials are cut from file-folder material and attached with sewing snaps to a sheet of acetate. Once the problem is discussed and solved, move the gauge and capacity dials to vary the values in the problem. Then let the students modify the problem as well to include additional information:

How much farther can the car travel if it gets 22 miles to the gallon?

How much will it cost to fill the tank at 87.9¢ per gallon?

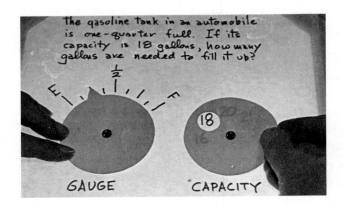

5.9 CALCULATORS AND COMPUTERS APPLICATIONS

The current age of technology has made the microcomputer available to virtually every school and the calculator to essentially every interested student. These electronic marvels are certain to have a long-term effect on the mathematics curriculum. Even at this point we need to reevaluate certain extensively taught computational algorithms in light of their current practical value. Calculators should be more readily available and their use more frequently encouraged. Students need experience in recognizing when a problem is most appropriately done mentally, with pencil and paper, or with a calculator. Problem-solving situations once avoided because of the tediousness of the necessary computations can now be explored with calculator and computer. Problem-solving strategies such as guess-and-test are now more accessible, and numerical tables for discovering patterns and critical values are now more easily generated.

The ideas that follow are but a brief sampling of the many possibilities that exist for classroom use of these new teaching tools. They relate to the arithmetic aspects of the curriculum but have objectives that extend well beyond those of just computing the correct answer.

Primes

To test to see if a number is prime, check for possible prime divisors up to the square root of the number. This example is illustrated using a calculator. A microcomputer program that tests a number for primeness is given on page 145.

1. Try 677. The calculator gives $\sqrt{677}$ as 26.019223. So the only possible prime factors are 2, 3, 5, 7, 11, 13, 17, 19, and 23. The first three factors can be eliminated immediately. Do any of the others give whole number quotients? Is 677 prime or composite?
2. What is the next prime number after 677?

Repeating Decimals

Repeating decimals offer a rich source of activities for calculator use in the classroom.

Divide the numbers 1 through 5 by 9.	$\frac{1}{9} = 0.1111111$
Look for a pattern. Guess the results	$\frac{2}{9} = 0.2222222$
for dividends of 6 through 8.	$\frac{3}{9} = 0.3333333$
What about 9?	$\frac{4}{9} = 0.4444444$
	$\frac{5}{9} = 0.5555555$

Now have each student guess these results as decimals using the calculator as a check.

100 divided by 9
2000 divided by 9
3,000,000 divided by 9

1. Use your calculator to explore the repeating decimals formed by dividing by 99 and by 999.
2. What patterns can you discover for dividing by 11?
   See page 149 for a microcomputer program that will list any number of digits in a repeating decimal expansion.

Pi

Many rational approximations to π can be found in recorded history.

Egypt	1800BC	$(4/3)^4$
Greece	240BC	3 1/7
Greece	150AD	377/120
China	470AD	355/113
India	510AD	62,832/20,000

Use a calculator to compare their decimal values to 3.1415926.

Students need to experiment with what can be done on most simple four-function calculators. Here are several illustrations:

Enter a number, press the multiplication key, and then repeatedly press the equals key. Are successive powers of the number displayed?

Enter a number, press the division key, and then the equals key. Is the reciprocal of the number displayed? As you repeatedly press the equals key, are successive powers of the reciprocal displayed?

The first Student Worksheet that follows gives practice in finding powers of 4 and their reciprocals. Encourage students to use the techniques explored above. It then brings these two ideas together in looking for a discovery about the limiting value of the sums of successive reciprocals by making use of the memory-plus key.

Worksheet 2 uses the calculator to explore an interesting convergence problem involving fractions. Start with a fraction $\frac{a}{b}$ between 0 and 1. Add the numerator and denominator to form a new denominator and add this sum to the original denominator to form a new numerator.

$$\frac{a}{b} \rightarrow \frac{(a + b) + b}{a + b}$$

As the process is repeated, each time starting with the newly formed fraction, one discovers that the decimal equivalences appear to converge to the decimal for $\sqrt{2}$.

Powers, reciprocals, and sums

1. Find the first ten powers of 4, then find their reciprocals. If you add each reciprocal to memory as you go along, you can also find the successive sums of these reciprocals. Enter the reciprocals and their accumulated sums in the table.

EXPONENT ON 4	POWER OF 4	RECIPROCAL	RECIPROCAL SUM
1			
2			
3			
4			
5			
6			
7			
8			
9			
10			

2. Look at the decimal values in the last column. What simple unit fraction do the decimals appear to be approaching as more and more reciprocals of 4 are being added?

3. Follow the same procedure again but this time use reciprocals of successive powers of 3. What simple unit fraction do the decimals in the last column appear to approach as more and more reciprocals of powers of 3 are being added?

4. The answers to questions 2 and 3 can lead you to an interesting generalization. Can you discover it? If you can, you should have no trouble guessing the sum of the reciprocals of successive powers of 2? of 5? of n?

Student Worksheet 2

Converging ratios

Start with a fraction between 0 and 1.
Follow these steps to form a new fraction.

1. New denominator—Add the numerator and denominator of the original fraction.
   New numerator—Add this new denominator to the original one.
2. Write the new fraction and use your calculator to find a decimal equivalent to four decimal places.
3. Repeat the steps again, this time starting with the new fraction. Continue the process until you make a discovery about the decimal equivalent.

STARTING FRACTION _____

NEW DENOMINATOR	NEW NUMERATOR	NEW FRACTION	DECIMAL EQUIVALENT

4. Now try the same process with another fraction between 0 and 1.

Many of the topics of number theory lend themselves to microcomputer applications. Several such programs are presented here. They can be modified as desired to treat the topics at a more intuitive or a more sophisticated level.

Prime number test

```
5    REM  PRIME NUMBER TEST
10   INPUT "ENTER A POSITIVE INTEGER ";N
20   IF N < 1 THEN 10
30   IF N = 1 THEN 110
40   IF N = 2 OR N = 3 THEN 90
50   IF N / 2 = INT (N / 2) THEN 130
60   FOR K = 3 TO SQR (N) STEP 2
70   IF N / K = INT (N / K) THEN 130
80   NEXT K
90   PRINT N" IS PRIME"
100    GOTO 999
110    PRINT N" IS NEITHER PRIME NOR COMPOSITE"
120    GOTO 999
130    PRINT N" IS COMPOSITE"
999    END

JRUN
ENTER A POSITIVE INTEGER 243
243 IS COMPOSITE

JRUN
ENTER A POSITIVE INTEGER 257
257 IS PRIME
```

Greatest common factor

```
5    REM  GREATEST COMMON FACTOR
10   INPUT A,B
15   IF A < 1 OR B < 1 THEN 10
20   IF A > B THEN C = B: GOTO 40
30   LET C = A
40   FOR I = C TO 1 STEP - 1
50   IF I = 1 THEN 80
60   IF B / I = INT (B / I) AND A / I = INT (A / I) THEN 90
70   NEXT I
80   PRINT "THE NUMBERS ARE RELATIVELY PRIME"
90   PRINT "THE GCF IS ";I
999    END
```

```
JRUN                      JRUN
?952,391                  ?157,83
THE GCF IS 17             THE NUMBERS ARE RELATIVELY PRIME
                          THE GCF IS 1
```

Listing prime numbers

```
5    REM  LIST PRIME NUMBERS
10   PRINT "PRIME NUMBERS"
15   INPUT "BETWEEN ";A
16   INPUT "AND ";B
17   PRINT
20   FOR N = A TO B
30   FOR K = 1 TO N
40   IF N / K = INT (N / K) AND
     N / K > 1 AND N / K < N THEN 70
50   NEXT K
60   PRINT N;"   ";
70   NEXT N
999  END
```

```
]RUN
PRIME NUMBERS
BETWEEN 2
AND 100

2   3     5   7  11  13  17  19  23
   29    31    37  41  43  47  53  59
   61    67    71  73  79  83  89  97

]RUN
PRIME NUMBERS
BETWEEN 350
AND 375

353  359  367  373
```

Perfect numbers

This program is written to find all perfect numbers less than 100. A *perfect number* is an integer that is equal to the sum of all of its factors except for the number itself. The program calls for a very large number of divisions. In testing 2, it uses only one trial divisor. In testing each number N from 3 through 99, it uses N-2 trial divisors, D. Thus the total number of divisions required is 4754.

$$1 + (1 + 2 + 3 + \ldots + 95 + 96 + 97) = 1 + \frac{97(98)}{2} = 4754$$

Have your students first determine the number of divisions required in the program using the method shown above. Then have them time the running of the program to see how long it takes. Students need to appreciate the speed at which the microcomputer can do repeated computations.

```
10   PRINT "PERFECT NUMBERS LESS THAN 100"
20   FOR N = 2 TO 99
30   LET S = 1
40   FOR D = 2 TO N - 1
50   IF N / D < > INT (N / D) THEN 70
60   LET S = S + D
70   NEXT D
80   IF S < > N THEN 100
90   PRINT N
100   NEXT N
999   END

]RUN
PERFECT NUMBERS LESS THAN 100
6
28
```

While the speed of the microcomputer may be impressive, the scope of the problem is even more so. The next perfect number after 6 and 28 is 496. Increasing this program to search for all perfect numbers less than 500 would extend the computing time very significantly. However, testing a selected interval, such as 475 to 500, would reduce this required time substantially. Over the recent years, an extraordinary amount of computer time on large high-speed mainframes has been spent on exploring perfect numbers. At this point, some 30 perfect numbers have been found, most enormously large in magnitude. No one yet knows of an easy way to find them nor how many there actually are. The first nine perfect numbers are listed here.

$$6$$
$$28$$
$$496$$
$$8128$$
$$33,550,336$$
$$8,589,869,056$$
$$137,438,691,328$$
$$2,305,843,008,139,952,128$$
$$2,658,455,991,569,831,744,654,692,615,953,842,176$$

The eighteenth perfect number has 1937 digits.
The twenty-seventh perfect number has 26,790 digits!

Figurate numbers

Figurate numbers were first discussed on page 120 in this chapter. The microcomputer can be a handy tool for generating tables of figurate numbers. This program in BASIC lists the first 25 triangular, square, and pentagonal numbers.

```
5     REM   TRIANGULAR, SQUARE, AND PENTAGONAL NUMBERS
10    PRINT "FIGURATE NUMBERS"
20    PRINT
30    PRINT "TRIANGULAR      SQUARE      PENTAGONAL"
40    FOR N = 1 TO 25
50    LET T = N * (N + 1) / 2
60    LET S = N * N
70    LET P = N * (3 * N - 1) / 2
80    PRINT T,S,P
90    NEXT N
999   END


]RUN
FIGURATE NUMBERS
```

TRIANGULAR	SQUARE	PENTAGONAL
1	1	1
3	4	5
6	9	12
10	16	22
15	25	35
21	36	51
28	49	70
36	64	92
45	81	117
55	100	145
66	121	176
78	144	210
91	169	247
105	196	287
120	225	330
136	256	376
153	289	425
171	324	477
190	361	532
210	400	590
231	441	651
253	484	715
276	529	782
300	576	852
325	625	925

The formula used in line 60 of the program can also be written algebraically as $N(2N)/2$. In this form a very nice pattern can be found among the three formulas. Can you spot it and extend it to hexagonal, heptagonal, and octagonal numbers?

$$T = N(N + 1)/2 \qquad S = N(2N)/2 \qquad P = N(3N - 1)/2$$

Suppose triangular, square, and pentagonal numbers are said to be figurate numbers of order 3, 4, and 5 respectively. Then a figurate number of order k would be of the form

$$K = N[(K - 2)N + (4 - K)]/2$$

Repeating decimals

While the calculator can be a valuable aid when teaching fractions and their repeating decimal equivalences, the microcomputer can enhance the study even further. This program prints out the first 36 digits in the decimal expansion, allowing for a more detailed analysis of the repeating patterns in the digits. The runs given for $\frac{1}{7}$, $\frac{3}{19}$, and $\frac{4}{23}$ show decimals repeating every 6, 18, and 22 digits respectively. The program can easily be changed to print 100 or more successive digits.

```
5    REM   REPEATING DECIMALS
10   INPUT "NUMERATOR ";N
20   INPUT "DENOMINATOR ";D
30   LET A = INT (N / D)
40   PRINT A;".";
50   FOR C = 1 TO 36
60   LET N = (N - (A * D)) * 10
70   LET A = INT (N / D)
80   PRINT A;
90   NEXT C
100   PRINT ".."
999   END

]RUN
NUMERATOR 1
DENOMINATOR 7
0.142857142857142857142857142857142857..

]RUN
NUMERATOR 3
DENOMINATOR 19
0.157894736842105263157894736842105263..

]RUN
NUMERATOR 4
DENOMINATOR 23
0.173913043478260869565217391304347826..
```

Compound interest

The microcomputer can be especially useful as a classroom tool for producing lists and tables. Reading and analyzing these lists and tables can be a valuable problem-solving experience for the students as illustrated with this compound interest program.

Various deposits and annual interest rates can be entered and the amounts at different years examined when compounded annually and quarterly. The program can easily be modified to compound monthly and even daily.

```
5    REM  COMPOUND INTEREST
10     PRINT
20     INPUT "DEPOSIT ";D
30     INPUT "ANNUAL RATE ";R
40     INPUT "NUMBER OF YEARS ";N
50     PRINT
60     PRINT "YEAR","COMPOUNDED"
70     PRINT ,"ANNUALLY    QUARTERLY"
80     PRINT
90     PRINT "0",D,D
100    LET A = D
110    LET Q = D
120    FOR Y = 1 TO N
130    PRINT Y,
140    LET A = A + A * R
150    LET AR = INT (A * 100 + .5) / 100
160    PRINT AR,
170    FOR M = 1 TO 4
180    LET Q = Q + Q * R / 4
190    NEXT M
200    LET QR = INT (Q * 100 + .5) / 100
210    PRINT QR
220    NEXT Y
999    END
```

For a given time, students can compute how much more is earned when compounded quarterly rather than annually.

For a given deposit, students can compare earnings at different interest rates.

For a given rate, students can find when the deposit will double or triple in value.

This run shows that at $7\frac{1}{2}$% per year, money will double in 10 years compounded annually and quarterly. The money will triple in 15 years compounded quarterly.

```
]RUN

DEPOSIT 100
ANNUAL RATE .075
NUMBER OF YEARS 15

YEAR           COMPOUNDED
               ANNUALLY    QUARTERLY

0              100          100
1              107.5        107.71
2              115.56       116.02
3              124.23       124.97
4              133.55       134.61
5              143.56       144.99
6              154.33       156.18
7              165.9        168.23
8              178.35       181.2
9              191.72       195.18
10             206.1        210.23
11             221.56       226.45
12             238.18       243.92
13             256.04       262.73
14             275.24       283
15             295.89       304.83
```

EXERCISES

1. The four digits 5, 6, 7, and 8 are placed on separate cards. How many counting numbers can be formed using one or more of the cards?

2. The four digits 6, 7, 8, and 9 are used to form two two-digit numbers. Which two have the greatest difference? the greatest product?

3. The four numbers 4, 5, 6, and 7 are used as numerators and denominators for two fractions. Which two have the greatest sum? the greatest difference?

4. Represent the numbers from 1 through 20 using the digits of the current year. Any familiar mathematical operation may be used. Try to use the digits in the order in which they appear in the year.

5. Extend the list using four 4's on page 113 to represent as many of the numbers from 11 through 20 as possible.

6. Find several ways to represent 1000 using all nine digits from 1 through 9. (For example, $291 + 678 + 35 - 4 = 1000$)

7. Is a 10% raise followed by a 10% cut in pay better, worse, or the same as a 10% cut followed by a 10% raise? Explain why neither yields the original salary.

8. Write the six expressions that can be represented using a flow chart with an input of 6 and the three operations of "add 8," "multiply by 3," and "square."

9. For a given input value, a flow chart has four successive operations. In how many different ways can the operations be arranged? Explain why all the different arrangements may not necessarily produce different output values.

10. Order the three decimals $0.2\overline{33}$, $0.\overline{2332}$, and $0.2\overline{332}$. Then write a decimal for a rational number that lies between the smaller two and an irrational number that lies between the larger two.

11. What is the one millionth digit in the decimal expansion of $\frac{1}{7}$?

12. What is the last digit in the decimal expansion of $\frac{1}{8}$?

13. How many digits are there before the one hundredth 3 in the nonrepeating decimal 0.43443444344443444443 . . . ?

14. Write down any three-digit number and then repeat the same digits again to form a six-digit number. Now divide by 7, then divide the quotient by 11, and finally divide the resulting quotient by 13. Repeat the same procedure, starting with another three-digit number. State your discovery, and then provide a mathematical explanation as to why the trick works as it does.

15. Write down the first eight hexagonal figurate numbers. What is the nth hexagonal number?

16. Construct a 7 x 7 magic square using consecutive integers starting with 5.

17. Construct a 5 x 5 magic square using multiples of 3 starting with 3. What is the sum of each row, column, and diagonal? What is the middle entry?

18. What is the sum of all entries in an nth-order magic square formed from consecutive integers starting with 5?

19. Show that the sum of each row, column, and diagonal in an nth-order magic square formed from successive integers starting with 1 is $n(n^2 + 1)/2$. Explain why it follows that the middle entry must be $(n^2 + 1)/2$.

20. Show that the maximum number of regions that can be formed in a plane by a circle with n secants is $\dfrac{(n + 2)(n + 3)}{2} - 2$.

21. Prove that the sum of the first n odd counting numbers is n^2.

22. Prove that the "spider" nomograph on page 137 produces the relationship $\dfrac{1}{a} + \dfrac{1}{b} = \dfrac{1}{k}$.

ACTIVITIES

1. Mark twenty cards with the numbers 1 through 20. Shuffle them and deal out the top five cards. Use each of the selected values once, with any of the four fundamental operations to give the value on the sixth card. Practice by playing the game with a friend.

2. Prepare several secret message charts such as the one shown on page 114. Have one of these feature computation with fractions and another computation with decimals.

3. Start with a rectangular sheet of paper. Fold the left edge over to the right edge and then reopen. Refold repeatedly, each time folding the left edge to the crease just made and then reopen. Set up a table of values relating the total number of rectangles formed to the total number of folds made. Generalize the number pattern for n folds.

4. Fold a paper square in half vertically and horizontally. Without unfolding, fold it in half again vertically and horizontally. Guess at the total number of rectangles of all sizes before you open up the square and count.

5. Describe how you might lead a class to discover the answer to question 4 by looking for a pattern using some simpler cases.

6. Make up a variety of games on mathematical skills that can be played with a simple deck of ten 3 x 5 file cards numbered 0 through 9.

7. Describe how you would explain the difference between "200% of 24" and "200% more than 24" to an eighth-grade class.

8. Prepare a set of 16 punched cards that can be sorted in just four operations. Then describe how, once shuffled, they can be sorted to spell out the message MATHEMATICAL FUN.

9. Construct a laboratory worksheet that can be used to lead a student to discover that the sum of each row, column, and diagonal in an nth-order magic square formed from the integers 1 through n is $n(n^2 + 1)/2$.

10. Prepare a short lesson plan on how you would show an eighth-grade class how to construct various 3 x 3 magic squares. Include three examples, one with fractions, one with decimals, and one with integers.

11. Construct a magic square on a model of a torus as described in Martin Gardner's Second Scientific American Book of Puzzles and Diversions. New York: Simon and Schuster, 1961.

12. Do research on the interest in magic squares shown by the statesman Benjamin Franklin and the artist Albrecht Dürer.

13. Construct a model of a 3 x 3 x 3 magic cube as described on page 129.

14. Prepare a lesson plan on adding integers that makes use of student-constructed nomographs.

15. Construct a percent nomograph from strips of one-cycle and two-cycle semilog graph paper.

16. Make a flow chart transparency similar to those described on page 134 that can be used to practice mental computation with whole numbers.

17. Describe a lesson that would make use of flow charts in developing the role of parentheses in arithmetic expressions.

18. Prepare a flow chart transparency that can be used to review skills in percent. List various inputs and operations on movable strips of acetate.

19. Prepare the percent chart described on page 111 on a transparency and discuss how it might be used in teaching a lesson on percents.

20. Construct a transparency for reviewing powers. Place the different values of the exponent and base on movable acetate dials as shown on page 140.

21. Complete the calculator activity on powers, reciprocals, and sums described on Worksheet 1, page 143.

22. Write a program in BASIC that can be used to generate the first ten decagonal numbers.

READINGS AND REFERENCES

1. The 1986 Yearbook of the National Council of Teachers of Mathematics is Estimation and Mental Computation. Prepare a written summary of two of its chapters.

2. Read The Lore of Large Numbers by Philip Davis, Washington, D.C.: The Mathematical Association of America, 1961. Summarize two specific topics of

interest from Part I and two from Part II that would be especially useful in teaching mathematics at the secondary level.

3. For many years, Martin Gardner wrote on mathematical recreations in *Scientific American*. Numerous collections of these mathematical puzzles, paradoxes, diversions, and games have now been published in book form. Locate one such book, identify it by name, and summarize one of the chapters that deals with numbers.

4. One very useful reference for the mathematics classroom is an almanac. Identify a specific current almanac by name. Describe how three different mathematical topics could be presented using data found in the reference. Include the data in the report.

5. One of the classics on recreational mathematics is *Mathematical Recreations* by Maurice Kraitchik, New York: Dover Publications, Inc., 1953. Read his chapter 7 on magic squares and summarize several types of special squares mentioned.

6. Read Chapter 14 in *Men of Mathematics* by E. T. Bell, New York: Simon and Schuster, 1965. Prepare a report on the mathematician Bell calls "The Prince of Mathematicians."

7. Prepare a report on three capsules from Part I, The History of Numbers and Numerals, and three from Part II, The History of Computation, found in the 31st Yearbook of the National Council of Teachers of Mathematics titled *Historical Topics for the Mathematics Classroom*, 1969.

8. Read Chapter 2, "Beyond the Googol," from *Mathematics and the Imagination* by Edward Kasner and James Newman, New York: Simon and Schuster, 1967. Summarize how several of the topics presented could be incorporated into lessons for the mathematics classroom.

9. Refer to a history of mathematics book and read about the gelosia algorithm for multiplication popular in the late 15th and 16th centuries. Compare this method to that used with Napier's Rods.

10. Locate and identify by name five specific references on prime numbers that would be suitable as source material for a high school student preparing a report on the topic.

11. Study two different junior high school mathematics textbook series. Compare the ways they utilize calculator activities and computer programming to illustrate numerical patterns and properties and computational procedures.

12. One component of the current junior high school curriculum is computer literacy. Review a computer literacy textbook for this level and outline the key components in BASIC that are presented. Then write a brief lesson plan on a mathematical topic that utilizes the microcomputer and programming by the teacher. Several appropriate references are listed here.

Computer Literacy—A Hands-On Approach by Arthur Luehrmann and Herbert Peckman, New York: McGraw-Hill Book Company, 1983.

Computer Literacy with an Introduction to BASIC Programming by Neal Golden, Orlando, Florida: Harcourt Brace Jovanovich, 1986.

Using Computers by Gerald Elgarten and Alfred Posamentier, Menlo Park, California: Addison-Wesley Publishing Company, 1984.

Classroom Aids and Activities: Algebra

Chapter 6

One often thinks of concrete models and aids in the early stages of learning mathematics at the elementary level. While their roles may change somewhat, these same experiences remain equally important in the middle, junior high, and senior high school years. The abstract nature of algebra makes manipulative experiments, visualization activities, and motivational introductions, reviews, and extensions all the more valuable. This chapter looks at the use of aids and activities in the algebra class.

6.1 MOTIVATING ALGEBRA

Variety is a key element in the effective teaching of algebra. Not only does it offer a change of pace, but it also provides opportunities for bringing aids and activities into the classroom.

Geometric Models

Geometric models often give a much needed concrete stepping-off point into an algebraic generalization as illustrated here.

Each set of stairs shown on the following page can be used to represent the sum of the numbers 1 through 6. Put two together and you get a rectangle measuring 6 by 7.

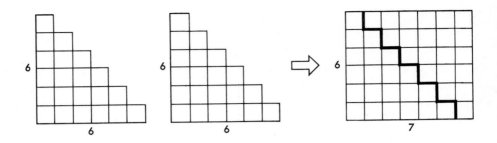

The rectangle has 6×7 or 42 squares, but it was made from two sums of 1 through 6. Hence,

$$1 + 2 + 3 + 4 + 5 + 6 = \frac{42}{2} = 21$$

Now imagine each set of stairs as having n steps. Two such sets joined together this way would form a rectangle measuring n by $n + 1$ with $n(n + 1)$ squares. It quickly follows that

$$1 + 2 + 3 + \cdots + n = \frac{n(n + 1)}{2}$$

The initial geometric model sets the stage for the transition to the general case for the variable n.

Proof Without Words

Algebraic proofs without words can be fun in the classroom. What algebraic idea can be discovered from these figures?

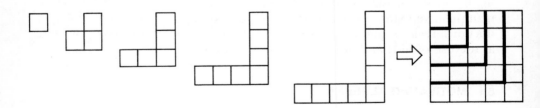

The figures illustrate an example of the generalization that the sum of the first n odd counting numbers is n^2.

Varied Format

For many, the notational manipulations of algebra need to be explained in more than one way. A common approach to multiplying binomials is shown on the left, but another method, given on the right, may be easier for some to understand.

These are just two different formats for the same process. They both give the same trinomial result, $8x^2 + 2x - 15$, but one more clearly shows that four partial products are needed.

Algebraic Analysis

Many of the familiar numerical curiosities and puzzles designed to create, motivate, and maintain interest have algebraic analyses that help to show the student just how useful the tools of algebra can be. Consider this little mathematical trick involving the number 1089.

STEPS	EXAMPLE
Write any three-digit number such that the hundreds digit is at least two more than the units digit.	Choose 783.
Reverse the digits and subtract.	783 − 387 396
Reverse the digits again and add.	396 + 693 1089

The surprising result is that the answer will always be 1089! Clever teachers can make this property work for them in many ways. One illustration is given here.

> Express your birth date as a four-digit number. Use the first two for the month and the last two for the day. Add it to the number you found in the last step above.

To guess the date, the teacher simply subtracts 1089. Every final result has a unique birth date. Here are some examples.

Final result	1206	1800	2106	2196
	− 1089	− 1089	− 1089	− 1089
	0117	0711	1017	1107
Birth date	Jan. 17	July 11	Oct. 17	Nov. 7

To the teacher, the number 1089 is special. But why does it always arise when you follow this process? An algebraic analysis gives the answer.

Let a be the hundreds digit,
  b be the tens digit, and
  c be the units digit, with $a > c + 1$

Original number	$100a + 10b + c$
Reverse digits	$100c + 10b + a$
Subtract	$100(a - c) + (c - a)$
	$= 100(a - c - 1) + 90 + (10 + c - a)$
Reverse digits	$100(10 + c - a) + 90 + (a - c - 1)$
Add	$100(9) + 180 + 9$
	$= 1089$

Tying the algebra to this numerical curiosity makes it a valuable topic for discussion in the classroom.

6.2 SOLVING EQUATIONS

A substantial proportion of the algebra curriculum centers around the solving of equations of various types. The solution of two-step linear equations can be especially troublesome for those who lack a meaningful model with which to associate the algorithmic procedures. This section is designed to offer some suggestions.

Flow Chart Format

Many students have difficulty solving an equation in algebra because they do not know what the equation says nor do they know how to find the particular sequence of steps needed to solve it. A simple, yet surprisingly effective technique for introducing this topic makes use of flow charts. The obvious step-by-step sequencing built into a flow chart is used to analyze the correct sequence needed to solve the corresponding equation.

Cut out a series of circular and rectangular pieces of paper for input/output and operations. Attach them to the board to form a flow chart. As an alternative, draw them on pieces of acetate and arrange them on an overhead projector. Those shown here will be used in the illustrations that follow, but they can obviously be varied and expanded. For each operation given, the corresponding inverse operation is needed.

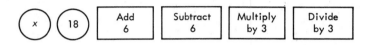

Form some flow charts using two of the available operation pieces. Have the students write the corresponding equations.

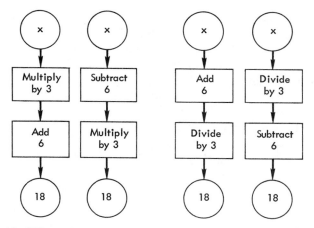

In all, 12 different two-step equations are possible from the operations shown using *x* as input. Which pairs of equations will have the same solutions and why?

Now write out some of the possible equations, such as these, and have the students correctly select and arrange the corresponding pieces for the flow chart.

$$3(x + 6) = 18 \qquad \frac{x - 6}{3} = 18$$

This activity gives the students a physical format in which to associate abstract equations. Substantial drill at this level can produce significantly better performance in the writing and understanding of equations and word problems. Once the students have this experience in putting equations together, ask how they can be taken apart. It is this basic idea, easily shown with flow charts, that leads to a concrete model for solving equations.

Start with a flow chart and the corresponding equation. Then, beginning with the output, reverse the sequence, performing the inverse operations as you go.

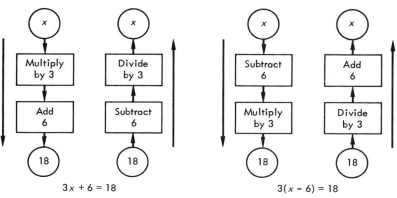

$$3x + 6 = 18 \qquad\qquad 3(x - 6) = 18$$

Flow chart pieces can be cut from acetate and arranged for projection using the overhead projector. Similarly, a two-step flow chart pattern can be cut from heavy paper and projected on the chalkboard. Input/output and operations can be written in and erased as desired while the projected flow chart outline remains.

For more advanced classes, construct, solve, and compare some three-step equations as well. These require two additional flow chart pieces, "multiply by 4," and "divide by 4."

$$\frac{3x + 6}{4} = 18 \qquad \frac{4x + 6}{3} = 18$$

$$\frac{4x}{3} + 6 = 18 \qquad \frac{3x}{4} + 6 = 18$$

Rule of False Position

Long before algebra as we know it was invented, the ancient Egyptians were solving equations using a strange "guess, test, adjust" process called the *rule of false position*.

As an example, consider solving this equation.

$$x + \frac{x}{7} = 24$$

First, choose a convenient guess, say 7.
Next, use 7 in place of x on the left.

$$7 + \frac{7}{7} = 8$$

Our guess gives a value of 8, but we wanted 24. Since 24 is 3 times 8, the solution must be 3 times the guess of 7. Check to see that 21 is the solution.

Here are two problems from the Rhind Papyrus, an ancient Egyptian scroll dating back to 1650 BC. Use the method of false position to solve each problem.

A quantity and its 1/7 added together become 19. What is the quantity?

A quantity and its 1/2 added together become 16. What is the quantity?

The *rule of double false position* is more complicated, but it can be used very effectively to introduce or motivate a unit on the solution of pairs of linear equations. Here one begins by taking two guesses, x_1 and x_2. These guesses are then substituted in the given equation and the results are noted as r_1 and r_2. The correct solution is then found by substitution in the following formula:

$$x = \frac{r_1 x_2 - r_2 x_1}{r_1 - r_2}$$

Consider the equation $3x - 12 = 0$ with guesses $x_1 = 2$ and $x_2 = 5$.

$$\text{For } x_1 = 2: r_1 = 3(2) - 12 = -6$$
$$\text{For } x_2 = 5: r_2 = 3(5) - 12 = 3$$

By substitution in the formula we have

$$x = \frac{(-6)(5) - (3)(2)}{-6 - 3} = \frac{-36}{-9} = 4$$

After applying the formula to several problems to be assured that it works, the class can be asked to justify it and can be led to do so by considering the general linear equation $ax + b = 0$.

$$\text{For } x = x_1: \quad \begin{cases} ax_1 + b = r_1 \\ ax_2 + b = r_2 \end{cases}$$
$$\text{For } x = x_2:$$

Now we also know that the solution of the general equation $ax + b = 0$ is $x = -b/a$. Thus all that is needed to verify the given formula for the rule of false position is to use the linear equations given and find $-b/a$ in terms of x_1, x_2, r_1, and r_2.

6.3 GRAPHING

René Descartes' contribution to mathematics was immense. In laying out the elements of coordinate geometry, he brought together algebra and geometry. Abstract algebraic expressions were given visual geometric counterparts that could be seen. Now an equation such as $y = 3x - 4$ appears as a specific straight line on a rectangular coordinate plane. Its graph can be drawn by the student. But not all students see the more general linear equation $y = ax + b$ as clearly.

Families of Lines

How do students view lines on a coordinate plane? Some see each and every possible line as a separate individual entity. Others see but a single general line, like a pencil lying on a paper, that can be moved about into any position desired. A study of families of lines helps focus on this second notion by shedding light on the roles a and b play in the linear equation $y = ax + b$. But even more important, it shows how a line can be moved around the plane through rotation and translation.

Have the students graph $y = 3x - 4$ and then place their pencils over the line.

Keeping the y-intercept fixed, move the pencil through the family of lines containing $(0, -4)$. One can view this as rotating the original line about a fixed point.

Keeping the slope fixed, move the pencil in a parallel fashion through the family of lines with slope 3. One can view this as translating the original line.

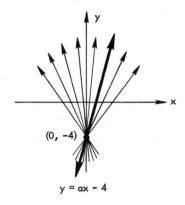

$y = ax - 4$

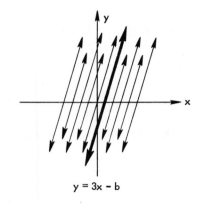

$y = 3x - b$

By rotating about the y-intercept and then translating, any line can be moved to any position. That is one way to think of the general linear equation $y = ax + b$.

How can any line be moved to any other nonparallel position in the plane by rotation alone? The answer lies in the fact that, since the two lines are not parallel, they must have a point in common. Use that point as the point of rotation.

Parabolas

There are many methods for drawing the graph of the parabola for $y = x^2$. This activity ties the graph to an interesting number property related to squares.

Start at the origin. Move one unit at a time to the right, moving up 1, then up 3, then up 5, up 7, and so on. Mark each point. Now repeat the process, moving one unit at a time to the left.

Draw a smooth curve through the points for the graph of $y = x^2$.

This method makes use of the relationship between successive squares and the sums of successive odd numbers as shown on the following page.

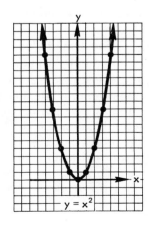

$y = x^2$

$$1 = 1 = 1^2$$
$$1 + 3 = 4 = 2^2$$
$$1 + 3 + 5 = 9 = 3^2$$
$$1 + 3 + 5 + 7 = 16 = 4^2$$
$$1 + 3 + 5 + 7 + 9 = 25 = 5^2$$

What simple modification in the process would produce a whole family of parabolas passing through the origin?

Checkers and Chess

A variety of novel applications of checkers and chess can attract attention in the mathematics class. An introduction or review of coordinate geometry can take on a new twist when presented through the games of checkers and chess.

As an example, what sequence of jumps can be made against black in this position on a checkerboard? Move from (2, 4) to (4, 6) to (2, 8), capturing men at (3, 5) and (3, 7).

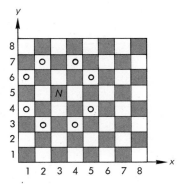

What move should be made if black moves next?

In chess, squares protected by various pieces can be given in coordinate form. What squares are protected by the knight at (3, 5)? The knight protects squares (1, 4), (1, 6), (2, 3), (2, 7), (4, 3), (4, 7), (5, 4), and (5, 6).

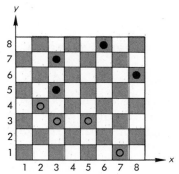

What unobstructed positions would a bishop at (7, 4) protect?

The concept of an equation as a set of number pairs satisfying a given condition can be reinforced by expressing the coverage of various chess pieces. The queen at (5, 6) protects unobstructed positions on these four lines.

$$x = 5 \qquad y = 6 \qquad y = x + 1 \qquad y = 11 - x$$

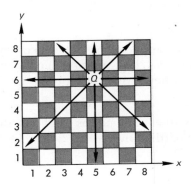

What lines would a rook at (2, 4) protect?

Variable Relationships

Students frequently lack concrete models that they can associate with familiar algebraic curves. Furthermore, they often fail to see these curves as geometric representations of variable relationships.

Tie the ends of a piece of string together to form a 22-inch loop. Show how rectangles of various lengths from 0 to 11 inches can be constructed from the string. Discuss how the width, perimeter, and area of the rectangle change as the length increases.

Following this discussion, ask the students to sketch the relationship between the length of the rectangle and its width, perimeter, and area.

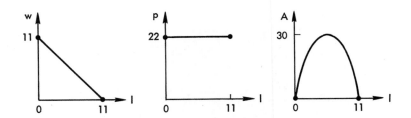

Complex Roots

Where are the roots of a quadratic equation when the discriminant is negative?

Complex roots in the form $x = a + bi$ have a real part a and an imaginary part bi. The model shown below illustrates these two components.

Make two grids on separate pieces of graph paper. On one grid draw the graph in the real plane where $b = 0$. On the other grid draw the graph in the imaginary plane where $a = 0$.

$$y = x^2 + 1$$

	x	y
on real plane ($b = 0$)	± 2	5
	±$\sqrt{3}$	4
	±$\sqrt{2}$	3
	± 1	2
	0	1
on imaginary plane ($a = 0$)	± 1i	0
	±$\sqrt{2}\,i$	−1
	±$\sqrt{3}\,i$	−2
	± 2i	−3
	±$\sqrt{5}\,i$	−4
	±$\sqrt{6}\,i$	−5

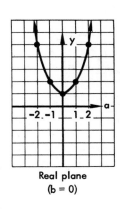

Real plane
(b = 0)

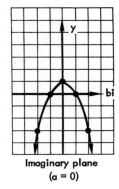

Imaginary plane
(a = 0)

Cut the two grids along opposite halves of the y-axis. Insert them together at right angles for an effective visual model showing the complex roots.

The model shown here was constructed on half-inch graph paper and can be made by the students. If you draw the grids on acetate, an impressive and especially visible model is formed.

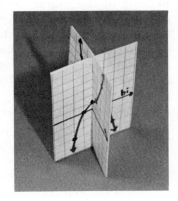

For the advanced algebra student, consider this circle.

$$x^2 + y^2 = 25$$

What does the curve look like for values of y greater than 5 and less than −5? The points are there but they are in the imaginary plane.

If $x = a + bi$, then
$$(a + bi)^2 + y^2 = 25$$
$$y^2 = 25 - (a + bi)^2$$
$$y^2 = 25 - a^2 - 2abi + b^2$$

Since y is real, $2ab = 0$. Hence, $b = 0$ or $a = 0$.

When $b = 0$, we have the circle in the real plane.	When $a = 0$, we have the hyperbola in the complex plane.

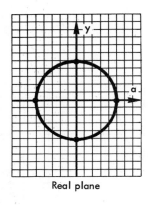

Real plane

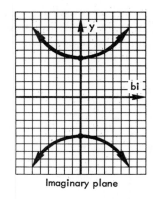

Imaginary plane

Draw these two graphs and assemble them as above to make a striking model of all the complex points, real and imaginary, for the equation $x^2 + y^2 = 25$.

6.4 SEQUENCES AND SERIES ACTIVITIES

Because of the undue emphasis placed on mechanics, the algebra class often lacks opportunities for student discovery. Nevertheless, much of the beauty of mathematics centers around the power of generalization. Since generalization is such an important problem-solving skill, it deserves frequent attention throughout the study of mathematics, but it can probably be developed best in an algebra course.

Paper Folding a Sequence

Here is a simple paper-folding activity that can generate some interest and lead into a discussion of infinite sequences and series.

Start with a unit square cut from a piece of paper.

Fold the four corners to the center.	The area is 1/2.
Fold the four corners in again.	The area is 1/4.
Repeat the process another time.	The area is 1/8.
Repeat again.	The area is 1/16.

1/2

1/4

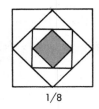

1/8

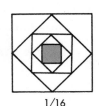

1/16

Use this activity to introduce geometric series.

Ask for the tenth term.   1/1024
Ask for the sum of the first ten terms.   1023/1024

Next generalize.

Ask for the nth term. $(1/2)^n$
Ask for the sum of the first n terms. $1 - (1/2)^n$

Now consider the infinite sequence.

Ask for the last term.   There is none.
Ask for the sum of all of the terms.   1

Fibonacci Sequence

In the Fibonacci sequence, each term after the second is the sum of the two preceding terms.

1, 1, 2, 3, 5, 8, 13, 21, 34, 55, 89, 144, . . .

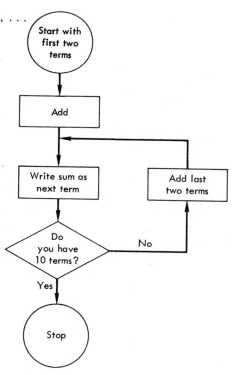

A flow chart offers a dynamic, visual format in which to present the rules for generating such a sequence. Cut the parts out and attach them to the chalkboard or bulletin board or use acetate and display them using the overhead projector.

A Fibonacci sequence can be formed by starting with any two numbers for the first two terms as noted in the input statement. Successive terms thereafter are always generated in the same way as noted by the operation statements.

In this flow chart, the decision box is used to control the number of terms to be generated. Obviously, each sequence could be continued without end.

One value the decision box plays for the teacher is that it can be changed to vary the nature of the problem and to enhance the problem-solving experience.

Start with the terms 1 and 1.
With the decision box shown above, the last term generated would be 55.

What would be the last term using each of these decision boxes instead?

Last term is 144   Last term is 377   No last term

What would the first 10 terms be if these were the starting terms?

1, 3, 4, 7, 11, 18, 29, 47, 76, 123   2, 5, 7, 12, 19, 31, 50, 81, 131, 212

The sum of the first 10 terms in any Fibonacci sequence can be used to illustrate an interesting property. Start by encouraging students to try to discover a connection between the seventh term and the sum of the first 10 terms. Have them test their conjectures with some additional Fibonacci sequences. Then let them try to prove algebraically that the property must hold. Try the process yourself before continuing.

This relationship affords the opportunity for an interesting classroom demonstration and a worthwhile lesson in discovering and establishing a generalization. The teacher plays the role of "magician" along the way, but it is the tools of algebra that yield the proof.

A student is asked to write any two numbers on the board. Assume they are 4 and 9. The class then helps to list the first 10 terms of a Fibonacci sequence that starts with these two numbers. When the class reaches the seventh term, 92, the teacher mentally multiplies this by 11 and writes 1012 on the side of the board. The class continues through the tenth term, and the students are then asked to find the sum. They are amazed to find that the sum is a number that the teacher had written down before they ever completed writing the sequence!

First	4	
Second	9	
Third	13	
Fourth	22	
Fifth	35	
Sixth	57	
Seventh	92	$11 \times 92 = 1012$
Eighth	149	
Ninth	241	
Tenth	390	
	1012	Sum of first 10 terms

Note that the sum of the first 10 terms is 1012, the same as 11 times the seventh term.

Finally the class is asked to try to discover the teacher's method. A hint may be given, if necessary, that the sum is a function of the seventh term. Possibly the class may need to be guided to represent the first 10 terms of a Fibonacci sequence in general terms, but this should be all the hint that is necessary.

First	a
Second	b
Third	$a + b$
Fourth	$a + 2b$
Fifth	$2a + 3b$
Sixth	$3a + 5b$
Seventh	$\boxed{5a + 8b} \Rightarrow$
Eighth	$8a + 13b$
Ninth	$13a + 21b$
Tenth	$\underline{21a + 34b}$
	$55a + 88b$ Sum of first 10 terms

Note that the sum of the first 10 terms is $55a + 88b$. By factoring, this is equivalent to $11(5a + 8b)$, which proves to be 11 times the seventh term.

Finite Differences

Some of the best classroom activities are the ones that start out with something concrete or geometric and lead into a number pattern, a conjecture, and a generalization established algebraically. When they also introduce new mathematics of significant importance, they are all the better. This activity illustrates the point by offering at the end a discussion of finite differences and a procedure for finding an nth-degree polynomial from $n + 1$ or more given data points.

What is the greatest number of regions into which a circle can be divided by x chords?

Begin by experimenting with some simple cases.

| 1 chord | 2 chords | 3 chords | 4 chords |
| 2 regions | 4 regions | 7 regions | 11 regions |

Close investigation reveals a connection between the number of regions

$$2, 4, 7, 11, 16, \ldots$$

and the successive sums of the counting numbers

$$1, 3, 6, 10, 15, \ldots$$

A generalization follows quickly.

Consider here another approach. Collect these data in a table and find the first and second differences.

Number of chords, x	Number of regions, y	First differences	Second differences
1	2		
2	4	2	1
3	7	3	1
4	11	4	

Inasmuch as the second differences are constant, we assume that the formula relating the number of regions (y) as a function of the number of chords (x) is given by the second-degree function $y = ax^2 + bx + c$. To solve for the constants a, b, and c we proceed to substitute the values given in the table and solve the three linear equations thus obtained.

$$\text{For } x = 1: \quad ax^2 + bx + c = a + b + c = 2$$
$$\text{For } x = 2: \quad ax^2 + bx + c = 4a + 2b + c = 4$$
$$\text{For } x = 3: \quad ax^2 + bx + c = 9a + 3b + c = 7$$

By subtracting the first equation from the second, and the second equation from the third, we obtain

$$3a + b = 2$$
$$5a + b = 3$$

Subtracting again gives $2a = 1$, or $a = \frac{1}{2}$. By substitution we find $b = \frac{1}{2}$ and $c = 1$.

The number of regions (y) is given in terms of the number of chords (x) by this equation:

$$y = \frac{1}{2}x^2 + \frac{1}{2}x + 1$$

Check the answer for $x = 4$; then verify with a drawing the answers for $x = 5$ and $x = 6$.

Another example of where this method of finite differences can be applied is the following:

The greatest number of pieces you can get with successive slices through a cake, starting with one slice, is

$$2, 4, 8, 15, 26, 42, 64, \ldots$$

What is the general term for x slices?

The third differences in this sequence are constant. Hence a cubic

$$y = ax^3 + bx^2 + cx + d$$

is needed to relate the maximum number of pieces (y) to the number of slices (x). Simultaneously solving the four equations for f(1), f(2), f(3), and f(4) yields the generalization

$$y = \frac{1}{6}x^3 + \frac{5}{6}x + 1$$

6.5 ALGEBRAIC MODELS TO MANIPULATE

The very essence and beauty of algebra lies in its structure and symbolism. Yet many students never fully understand the abstract concepts and algorithmic skills of algebra because they cannot see them in a real, physical sense. Aspects of algebra can be demonstrated with teaching aids that offer concrete, visual, geometric models for abstract algebraic ideas. Manipulatives can be touched, moved, and arranged for a new view, a fresh approach.

EXPERIMENT 1   Paper Folding for $(a + b)^2 = a^2 + 2ab + b^2$

Material

One square piece of paper per student.

Directions

1. Fold one edge over at a point E to form a vertical crease parallel to the edge. Label the longer and shorter dimensions a and b.

2. Fold the upper right-hand corner over onto the crease to locate point F. Folding this way, point F will be the same distance from the corner as point E.

3. Now fold a horizontal crease through F and label all outside dimensions.

4. Find the areas of the two squares formed. Find the areas of the two rectangles formed. Show that these four areas together must equal $(a + b)^2$.

Analysis

area of square Y: a^2
area of square X: b^2
area of rectangle W: ab
area of rectangle Z: ab

The original square paper measures $a + b$ on each edge and hence has an area of $(a + b)^2$. The four smaller parts can be combined to give an area of $a^2 + 2ab + b^2$. But their combined area must equal that of the original square. Therefore,

$$(a + b)^2 = a^2 + 2ab + b^2$$

EXPERIMENT 2   Paper Folding for $(a + b)(a - b) = a^2 - b^2$

Material
One rectangular sheet of paper per student.

Directions

1. Fold the left edge down onto the bottom edge to locate point P.

2. Fold vertically through point P. This forms a square on the left. Label the longer and shorter dimensions a and b.

3. Fold the upper right-hand corner over on the crease to locate point Q.

4. Now fold a horizontal crease through Q and label all outside dimensions.

5. Note the three rectangular and one square region formed.
   Show that the areas of Y and Z together equal both $(a + b)(a - b)$ and $a^2 - b^2$.

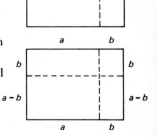

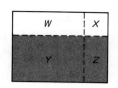

area of square X: b^2
area of rectangle W: ab
area of rectangle Y: $a(a - b) = a^2 - ab$
area of rectangle Z: $b(a - b) = ab - b^2$

Arrange pieces X, Y, and Z to form a square with area a^2.

Removing piece X gives an area of $a^2 - b^2$. But pieces Y and Z can be rearranged to form a rectangle with area $(a + b)(a - b)$. Hence,

$$a^2 - b^2 = (a + b)(a - b)$$

The two experiments just given can be modified to be more suitable for classroom demonstration by the teacher. Using scissors, the pieces can be cut, moved, and rearranged in a more visual way. This can be especially effective when done on the overhead projector.

EXPERIMENT 3   Cutting Up Squares

Material

Paper squares and scissors

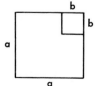

(a)

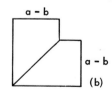

(b)

Directions

1. Cut a small square with sides b from a larger square with sides a.
2. Cut the remaining piece in half as shown.
3. Rearrange the two pieces to form a rectangle. What algebraic property is illustrated?

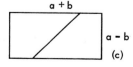

(c)

Analysis

Cutting the small square from the large one left an area of $a^2 - b^2$. Cutting this into two pieces and rearranging them, gives a rectangle with an area of $(a + b)(a - b)$. Since the pieces were only rearranged, the area remains the same. Hence,

$$a^2 - b^2 = (a + b)(a - b)$$

EXPERIMENT 4 Factoring a Trinomial

Material

A set of large squares measuring x by x, labeled by their areas as $x \times x$ or simply x^2.

A set of small squares measuring 1 by 1, labeled by their areas as 1×1 or simply 1.

A set of rectangles measuring x by 1, labeled by their areas as $x \times 1$ or simply x.

Directions

Monomials, binomials, and trinomials in x can now be represented by the appropriate geometric figures.

In factoring a trinomial, the various monomial parts are arranged in a rectangular shape. The dimensions of the rectangle give the factors.

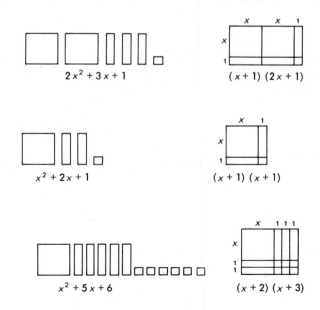

1. Show how to factor $3x^2 + 4x + 1$ using a model.
2. Show how to factor $4x^2 + 8x + 3$ using a model.

3. Show some other trinomials that can be factored using a model. Show some that cannot be factored.

4. Imagine a rectangular array of these pieces measuring $5x + 1$ by $2x + 3$. How many pieces are in it? How many of each size are there? What trinomial factorization does it represent?

Analysis

Consider the trinomial $ax^2 + bx + c$. See if a of the larger squares, b of the rectangles, and c of the smaller squares can be arranged to form a rectangle. If they can, then the dimensions of the rectangle give the two factors.

For $3x^2 + 4x + 1$, use 3 large squares, 4 rectangles, and 1 small square to get the factored form $(3x + 1)(x + 1)$.

For $4x^2 + 8x + 3$, use 4 large squares, 8 rectangles, and 1 small square to get a factored form of $(2x + 1)(2x + 3)$.

A rectangular array of $5x + 1$ by $2x + 3$ would require 30 pieces from the trinomial $10x^2 + 17x + 3$.

While the multiplication of two binomials usually does not present the difficulties that factoring does, it, too, can be introduced by a similar method. Draw a rectangle with dimensions those of the two binomial factors given. Then subdivide the rectangle into x^2's, x's, and 1's to find the product.

$(x + 1)(x + 2) = x^2 + 3x + 2$ $(x + 2)(x + 2) = x^2 + 4x + 4$

Small colored acetate figures on an overhead projector can be used instead of paper figures. Students can also try drawing the appropriate figures on graph paper at their seats.

Obvious restrictions are implied in this aid. Positive integral values are needed and only reasonably small numerical coefficients are practical.

As a possible extension, consider a trinomial with negative coefficients. Areas that represent positive coefficients are cut from white paper and those that represent negative coefficients are cut from red paper. Now the factoring of trinomials with both positive and negative coefficients can be illustrated. Negative values add out with positive values, so red areas are placed on top of white areas. Equal positive and negative areas can be added where needed.

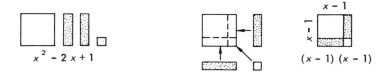

$x^2 - 2x + 1$ $(x - 1)(x - 1)$

How can the factorization of $x^2 + x - 2$ be shown using a model?

The experiments thus far have dealt with second-degree expressions and two-dimensional models. Experiment 5 presents a third-degree expression using a three-dimensional model that can easily be constructed in a school woodworking shop.

EXPERIMENT 5   A Model for $(a + b)^3$

Material

Eight blocks of wood cut to these dimensions, where a and b are any convenient lengths.

one	$a \times a \times a$
three	$a \times a \times b$
three	$a \times b \times b$
one	$b \times b \times b$

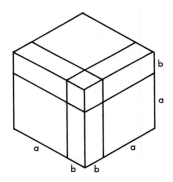

Directions

Arrange the eight pieces to form a cube. How does the resulting model illustrate this algebraic identity?

$$(a + b)^3 = a^3 + 3a^2b + 3ab^2 + b^3$$

Analysis

The cube measures $a + b$ on each edge so its volume is $(a + b)^3$. But it was made from eight pieces with this collected value.

$$a^3 + a^2b + a^2b + a^2b + ab^2 + ab^2 + ab^2 + b^3$$

Since the volumes of the original cube and the eight pieces combined are the same, we have

$$(a + b)^3 = a^3 + 3a^2b + 3ab^2 + b^3$$

6.6 MORE AIDS AND ACTIVITIES

Many believe that motivation is the very key to learning. An enthusiastic teacher can do much to stimulate, maintain, and increase interest in the subject by varying the things that are done in the classroom. This section contains more motivational ideas for the algebra classroom.

Mathemagic

This little trick can attract some real attention. Have all students participate.

STEPS	EXAMPLE
1. Begin by writing a three-digit number with all digits different.	257
2. Form all possible two-digit numbers from this number.	25 27 52 57 72 75
3. Find the sum of those two-digit numbers.	308
4. Divide the result by the sum of the three original digits.	$2 + 5 + 7 = 14$
The answer will always be 22!	$\begin{array}{r} 22 \\ 14\overline{)308} \\ 28 \\ \hline 28 \\ 28 \\ \hline \end{array}$

Here is why it works.

Let the number be abc with a value of $100a + 10b + c$.

There are six possible two-digit numbers.

$$10a + b$$
$$10a + c$$
$$10b + a$$
$$10b + c$$
$$10c + a$$
$$10c + b$$

The sum simplifies to 22 times $a + b + c$. So obviously, division by the sum of the digits must give 22.

$$20(a + b + c) + 2(a + b + c) = 22(a + b + c)$$

$$\frac{22(a + b + c)}{a + b + c} = 22$$

Introduce this trick and discuss it when studying factoring or the addition of polynomials.

Number Line Model

Students have little trouble recognizing that the whole numbers form an infinite set that goes on without end. Correspondingly, the number line extends without bound. But it is more difficult to accept the fact that there are just as many real numbers within a specific interval, say from 0 to 1. Yet every time students draw a continuous graph through selected points, they are using the completeness property of the real numbers and the corresponding concept of continuity of points on a line.

Have each student code his or her initials by replacing each letter with the appropriate digit or digits.

A	0	J	9	S	18
B	1	K	10	T	19
C	2	L	11	U	20
D	3	M	12	V	21
E	4	N	13	W	22
F	5	O	14	X	23
G	6	P	15	Y	24
H	7	Q	16	Z	25
I	8	R	17		

The resulting numbers will range up to six digits, to the hundred thousands. By placing a decimal point in front, they all become less than 1. Here are some examples.

L.P.M.	111512	0.111512
G.R.N.	61713	0.61713
C.A.F.	205	0.205
I.K.H.	8107	0.8107

Ask some questions at this point.

1. What three initials will code into the largest whole number possible?   ZZZ
2. What three initials will, when preceded by a decimal point, code into the largest decimal?   JJJ
3. What three initials will code into
   the smallest decimal greater than one-tenth?   KAB
   the largest decimal less than one-tenth?   AJJ
4. Is there a one-to-one correspondence between
   the initials and their coded whole numbers?   no
   the coded whole numbers and their decimals?   yes

Next, hold up a piece of string horizontally in front of the class as a model of the number line with your fingers locating the points 0 and 1. Have students come up and locate where the decimal codes for some familiar words such as these can be found.

algebra   geometry   mathematics

Then try the name of your town or state. Finally, try something like the Gettysburg address or the pledge of allegiance. Everything said and yet to be said—it is all there in that short interval from 0 to 1!

Babylonian Mathematics

A rich collection of interesting algebraic tidbits can be found in the annals of the history of mathematics. This one can be used when studying radicals or the Pythagorean theorem.

Pass out copies of this Babylonian clay tablet (1650 BC) to the students. Have them first try to guess what message it contains by studying the figure. Then offer the necessary clues for deciphering the cuneiform writing which represents numbers written in base 60. Finally, use a calculator to evaluate the fractions.

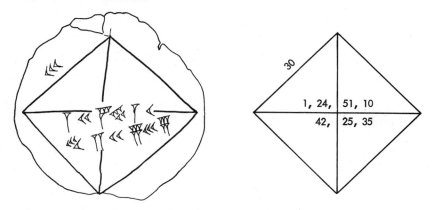

Interpreting the two sequences using these fractions, one can see both the method and precision of the ancient Babylonian computation. The diagonal of a 30-unit square is $30\sqrt{2}$ or $42.426405\dots$. Their result is amazingly close!

$$1, 24, 51, 10 = 1 + \frac{24}{60} + \frac{51}{60^2} + \frac{10}{60^3}$$
$$= 1 + 0.4 + 0.0141666 + 0.0000462$$
$$= 1.4142128$$

$$42, 25, 35 = 42 + \frac{25}{60} + \frac{35}{60^2}$$
$$= 42 + 0.4166666 + 0.0097222$$
$$= 42.4263888$$

All this happened more than 1000 years before the great Pythagoras lived!

Guessing a Quadratic

Here is a little mathemagical trick with an interesting algebraic analysis based on values of a quadratic equation. Its solution depends upon the mathematics of finite differences and, as such, is an excellent application for most students in their study of algebra.

EXAMPLE

Think of a quadratic function of the type $f(x) = ax^2 + bx + c$.

$f(x) = 3x^2 - 5x + 2$

Substitute 0, 1, and 2 for x in that order, and give the corresponding values for $f(x)$.

For $x = 0$, $f(0) = 2$
$x = 1$, $f(1) = 0$
$x = 2$, $f(2) = 4$

To determine the original expression, find the first and second differences.

$f(0), f(1), f(2)$

first differences

second differences

The coefficient of the x^2 term is one half of the number at the bottom; $\frac{1}{2}$ of $6 = 3$. The coefficient of the x term is the first number of the middle row minus one half of the number at the bottom ; $-2 - 3 = -5$. The first number of the top row is the constant, 2. Thus, the original expression is $3x^2 - 5x + 2$.

The explanation for these rules can be found by considering the general case, $f(x) = ax^2 + bx + c$. For this we have:

$$f(0) = c$$
$$f(1) = a + b + c$$
$$f(2) = 4a + 2b + c$$

Finding first and second differences we have the following:

$$c \quad a+b+c \quad 4a+2b+c$$
$$a+b \quad 3a+b$$
$$2a$$

From this array it is clear that one-half of the bottom number is a, the coefficient of x^2. The first number of the middle row minus one half of the bottom number gives $a + b - a = b$, the coefficient of x. The first number of the top row is c, the constant term.

6.7 OVERHEAD PROJECTOR IDEAS

The overhead projector can be used in a variety of ways in the algebra classroom. It can serve as a very powerful instructional tool when teaching graphing. It can be used to motivate a discovery which is then proved algebraically. It can also serve as a change of pace in the regular routine of algebraic drill and practice. Examples of these three are illustrated here. Many other applications are possible as well.

Teaching Graphing

Concepts in graphing can be dramatically and vividly illustrated with the overhead projector by using the graph of a function and the corresponding set of axes on different sheets of acetate. Here is an illustration in which absolute values are used. Begin by constructing two acetate sheets.

Sheet 1: A conventional set of axes in one color.
Sheet 2: The graph of the function $y = |x|$ in another color.

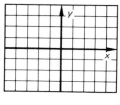

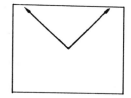

Sheet 1   Sheet 2

When sheet 2 is placed on top of sheet 1, the basic graph is projected.

$y = |x|$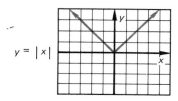

Displacing sheet 2 one unit up from its original position illustrates another graph. Displacing it two units down gives another.

$y = |x| + 1$ 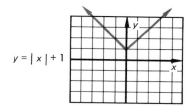 $y = |x| - 2$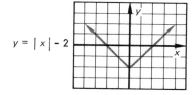

At this point other positions of vertical translation should be illustrated, with the class giving the proper equations. This first concept can then be further reinforced by offering the equation and having the student position the graph properly. The effect of vertical translation and the constant a in $y = |x| + a$ is thus illustrated.

Sheet 2 is now moved one unit to the right of its original position and the corresponding equation developed. Then it is moved two units left for another equation.

$y = |x - 1|$ 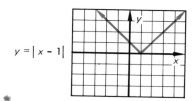 $y = |x + 2|$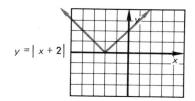

Again the horizontal translation is reviewed with additional examples by giving new equations and by giving new positions. The role of the constant b in $y = |x + b|$ is thus developed.

Now the two translations are combined for still other graphs.

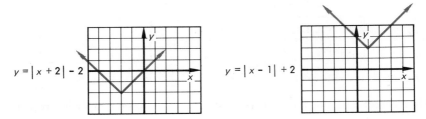

$y = |x + 2| - 2$ 　　　 $y = |x - 1| + 2$

These concepts can be reviewed with more positions and more equations until the student grasps the concepts involved in graphing equations of the form $y = |x + b| + a$.

When sheet 2 is flipped about the horizontal axis, a new graph is projected.

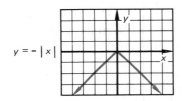

$y = -|x|$

Rotating 90° clockwise or counterclockwise illustrates two other graphs.

$$|y| = -x \qquad \text{and} \qquad |y| = x$$

A nice variation on this development can follow for reinforcement of concepts by placing sheet 1 on top of sheet 2 and moving the axes rather than the graph. The whole development is, of course, readily adaptable to other functions.

Motivating Discovery

In this illustration, a mask is used to isolate a selected square array of numbers from a given set. Students try to discover how to find the sum of the numbers in each array without adding them all. Then they try to prove their conjecture algebraically.

Step 1: The calendar for the month is copied on a sheet of acetate. Since accurate spacing is needed, use a piece of graph paper under the acetate for horizontal and vertical alignment of the numbers.

Step 2: With a razor blade a square is cut from a sheet of heavy paper or tagboard large enough to just expose a 3 × 3 array of numbers.

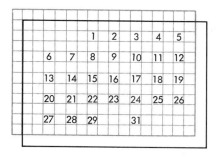

Step 3: By placing this mask over the number grid of the calendar, only a 3 × 3 array will be projected. Repositioning the mask simply exposes a new array.

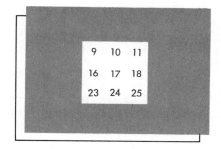

Students are asked to add the numbers in various arrays and then try to discover the relationship between each sum and the particular numbers exposed. Various suggestions should be encouraged and tested. The discovery hopefully will come from the students, aided, if necessary, with helpful hints. The sum of the numbers in each 3 × 3 array will always be 9 times the center number.

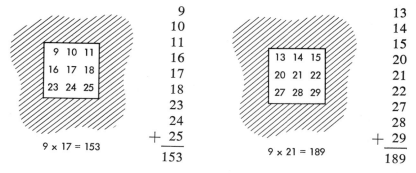

$9 \times 17 = 153$

$$\begin{array}{r} 9 \\ 10 \\ 11 \\ 16 \\ 17 \\ 18 \\ 23 \\ 24 \\ + 25 \\ \hline 153 \end{array}$$

$9 \times 21 = 189$

$$\begin{array}{r} 13 \\ 14 \\ 15 \\ 20 \\ 21 \\ 22 \\ 27 \\ 28 \\ + 29 \\ \hline 189 \end{array}$$

Encourage the students to verify the discovery algebraically. If the middle number is a, then the nine numbers can always be expressed this way regardless of where the mask is placed on the calendar. Illustrate this display with another transparency.

The sum of these numbers is obviously $9a$.

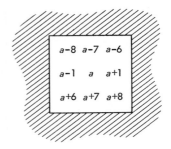

$a-8$	$a-7$	$a-6$
$a-1$	a	$a+1$
$a+6$	$a+7$	$a+8$

A clever variation of this idea involves the addition table.

0	1	2	3	4	5	6	7	8	9
1	2	3	4	5	6	7	8	9	10
2	3	4	5	6	7	8	9	10	11
3	4	5	6	7	8	9	10	11	12
4	5	6	7	8	9	10	11	12	13
5	6	7	8	9	10	11	12	13	14
6	7	8	9	10	11	12	13	14	15
7	8	9	10	11	12	13	14	15	16
8	9	10	11	12	13	14	15	16	17
9	10	11	12	13	14	15	16	17	18

Cut masks with 2×2, 3×3, 4×4, and 5×5 openings. Place these masks over the 10×10 addition table and see if the students can discover the relationship between the sum and the size of the array and the repeating number in the diagonal.

5	6
6	7

sum: $4 \times 6 = 24$

9	10	11
10	11	12
11	12	13

sum: $9 \times 11 = 99$

12	13	14	15
13	14	15	16
14	15	16	17
15	16	17	18

sum: $16 \times 15 = 240$

8	9	10	11	12
9	10	11	12	13
10	11	12	13	14
11	12	13	14	15
12	13	14	15	16

sum: $25 \times 12 = 300$

Once the discovery is made and verified, the mask can be removed from the number grid so that the entire 10×10 array is projected. Students can then compete to see who can find the total sum first.

$$100 \times 9 = 900$$

Now see how many students can generalize the procedure. For a similar square array going from 0 to n along the top row and down the left column, the lower right-hand number would be $2n$ and the total sum of all the numbers would be

$$n(n + 1)^2$$

Drill and Practice

Many algebraic skills call for frequent review and practice. If this can occur in a novel, unusual fashion, it will stimulate more interest. The following aid can be used to review binomial products. Factors can be easily changed so the speed in which the drill takes place can be controlled by the teacher. The level of difficulty can easily be adjusted by the proper selection of factors from among those available. And a wide choice of exercises is immediately available at your fingertips. This can be especially important when doing rapid-fire oral drill.

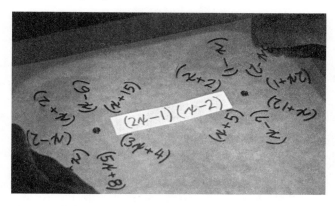

Mark two acetate disks with selected and varied binomial factors. Mount the two disks on the solid mask using sewing snaps so they can rotate easily. The teacher can see all the factors all the time but the class sees only one product projected at a time through the opening in the mask.

6.8 CALCULATOR AND COMPUTER APPLICATIONS

There are many possible uses of the calculator and computer in the algebra classroom. The study of convergent sequences and series offers many applications as does finding approximate roots of higher-degree equations.

Golden Ratio

The first 12 terms in the Fibonacci sequence are listed here.

$$1, 1, 2, 3, 5, 8, 13, 21, 34, 55, 89, 144$$

Have your students use a calculator to study the ratios of successive terms, larger to smaller. These results are from an 8-digit calculator.

1 to 1	1.0000000
2 to 1	2.0000000
3 to 2	1.5000000
5 to 3	1.6666666
8 to 5	1.6000000
13 to 8	1.6250000
21 to 13	1.6153846
34 to 21	1.6190476
55 to 34	1.6176470
89 to 55	1.6181818
144 to 89	1.6179775

The sequence of ratios is converging by oscillating below and above this limiting value called the *golden ratio*.

$$\frac{1 + \sqrt{5}}{2} = 1.6180339 \ldots$$

With even the greatest of patience and perseverance, the exact value can never be reached on a calculator. But interesting questions can still be asked and valuable experience gained on the concept of convergence.

Extend the Fibonacci sequence for another 12 terms. At what pair of adjacent terms is the ratio first correct to all the places on the calculator?

(4181 and 6765 for an 8-digit calculator)

The Number Pi

Many algebraic expressions have a value of π. This one, given by Leibniz, is especially interesting because of the simplicity of the pattern in the terms.

$$\pi = 4(1 - 1/3 + 1/5 - 1/7 + 1/9 - \ldots)$$

Ask the students to extend the pattern and evaluate the expression using the first 10 terms. Check to see how they use the calculator. By using reciprocals and adding to and subtracting from memory, the key sequence on most simple four-function calculators is extremely easy.

$$1 \div M^+ \quad 3 \div M^- \quad 5 \div M^+ \quad 7 \div M^- \quad 9 \div M^+ \quad 11 \div M^- \ldots$$

Just recall the accumulated value from memory and multiply by 4 after the desired number of terms are entered. Ask the class if the sequence is converging rapidly. The answer is an emphatic no! The fact is, most students would doubt the genius of Leibniz entirely were it based on this example.

To better understand the convergent property of this sequence, try this BASIC program where you input the number of terms desired.

```
10   PRINT
20   PRINT "LEIBNIZ'S SERIES FOR PI"
30   PRINT
40   INPUT "NUMBER OF TERMS ";N
50   LET A = 1
60   LET B = 1
70   LET PI = O
80   FOR C = 1 TO N
90   LET PI = PI + A / B
100  LET A = A * ( - 1)
110  LET B = B + 2
120  NEXT C
130  LET PI = 4 * PI
140  PRINT
150  PRINT "PI EQUALS ";PI
999  END
```

Try input values of 100, 1000, and 10,000. It will quickly become apparent that this series simply does not converge rapidly. But it does indeed converge in the limit to π.

Continued Fractions

Continued fractions give students a good opportunity to plan out the best sequence of steps to follow in solving a problem using the calculator. Encourage students to do all their work without writing down intermediate results.

$$1 + \cfrac{1}{1 + \cfrac{1}{1 + \cfrac{1}{1 + \cfrac{1}{1 + \ \ldots}}}}$$

On most simple four-function calculators, repeat this sequence of keys to evaluate the continued fraction. Work back from the bottom. The divide-equal keys take the reciprocal at each stage. Repeat as many times as desired, taking note of the display after each touch of the division key. Stop the process at the division key step.

$$\underbrace{1 \div \ = \ +}\ \underbrace{1 \div \ = \ +}\ \cdots$$

The simplicity of this continued fraction belies the fact that its limiting value is the golden ratio, $\dfrac{1 + \sqrt{5}}{2}$ or 1.6180339 . . .

As a follow-up, let students explore the following continued fraction using their calculators. The elegant tie displayed between the two integers 1 and 2 and the irrational $\sqrt{2}$ is another example of the striking beauty of mathematics.

$$\sqrt{2} = 1 + \cfrac{1}{2 + \cfrac{1}{2 + \cfrac{1}{2 + \cfrac{1}{2 + \ldots}}}}$$

This BASIC program shows how quickly this continued fraction converges. It stops printing when successive values in the machine are the same.

```
5    PRINT
10   PRINT "CONTINUED FRACTION"
20   PRINT "FOR SQUARE ROOT OF 2"
30   PRINT
40   PRINT 1
50   LET X = 2
60   LET X = 1 / X + 2
70   LET S1 = X - 1
80   IF S1 = S2 THEN 999
90   PRINT S1
100  LET S2 = S1
110  GOTO 60
999  END

]RUN

CONTINUED FRACTION
FOR SQUARE ROOT OF 2

1
1.5
1.4
1.41666667
1.4137931
1.41428572
1.41420118
1.41421569
1.4142132
1.41421363
1.41421355
1.41421356
1.41421356
```

The Number e

What happens to $(1 + 1/n)^n$ as n increases without bound? Get some reactions from your students on this question before continuing. The limit is the irrational, transcendental number e.

$$\lim_{n \to \infty} (1 + 1/n)^n = e = 2.7182818 \ldots$$

Have your students investigate this interesting number using a calculator and this series:

$$e = 1 + \frac{1}{1} + \frac{1}{2 \cdot 1} + \frac{1}{3 \cdot 2 \cdot 1} + \frac{1}{4 \cdot 3 \cdot 2 \cdot 1} + \frac{1}{5 \cdot 4 \cdot 3 \cdot 2 \cdot 1} + \cdots$$

Evaluate the first 10 terms of this series and find the sum. Compare the results with the value of e given above.

This continued fraction also has a value of e.

$$e = 1 + \cfrac{2}{1 + \cfrac{1}{6 + \cfrac{1}{10 + \cfrac{1}{14 + \cdots}}}}$$

Evaluate it for the values given using a calculator.

Roots of a Quadratic Equation

Planning the key parts desired in a computer program, writing out the steps in detail, running it, and verifying and debugging it are all important components of problem analysis in general. In that sense, programming assignments can help students organize and understand the facts that they learn.

Consider an assignment where students are asked to write a program that classifies the roots of the quadratic equation $AX^2 + BX + C = 0$.

```
5    REM    ROOTS OF A QUADRATIC
10   PRINT
20   INPUT "A = ";A
21   INPUT "B = ";B
22   INPUT "C = ";C
30   IF A = 0 THEN 20
40   LET D = B ^ 2 - 4 * A * C
50   IF D > = 0 THEN 70
60   PRINT "ROOTS ARE IMAGINARY."
65   GOTO 999
70   PRINT "ROOTS ARE REAL, ";
80   IF D = 0 THEN 120
90   IF  INT (1000000 *  SQR (D) + .5) / 1000000 =  INT ( SQR (D))
     THEN 110
100  PRINT "IRRATIONAL, AND UNEQUAL."
105  GOTO 130
110  PRINT "RATIONAL, AND UNEQUAL."
115  GOTO 130
120  PRINT "RATIONAL, AND EQUAL."
130  LET X1 = ( - B +  SQR (D)) / 2 * A
140  LET X2 = ( - B -  SQR (D)) / 2 * A
150  PRINT "X1 = "; INT (1000000 * X1 + .5) / 10000000,
160  PRINT "X2 = "; INT (1000000 * X2 + .5) / 10000000
999  END
```

By entering the values of A, B, and C, the roots are classified as real or imaginary, rational or irrational, and equal or unequal and their values printed when real. A good understanding of the role of the discriminant is needed as well as the nature and effects of rounding. Study the suggested program. It includes some details that may be unnecessary at one level of use. And yet at another level, it will not produce correct results. Can you find three values for A, B, and C where the program will produce roots inconsistent with the classifications it gives?

EXERCISES

1. Consider forming a flow chart with an input of x and an output of 24 with any two of the operations add 4, subtract 6, multiply by 8, and divide by 10. Write and then solve the 12 equations that can be represented.

2. Points are located for a parabola by starting at the origin and then moving two units at a time right and left successively, moving down 1, 3, 5, 7, and so on. What is the equation of the parabola formed?

3. Imagine a piece of string tied together in a 20-inch loop and formed into a rectangle. Sketch three graphs showing the relationship between all possible values of the length and the corresponding width, perimeter, and area. Then find the corresponding algebraic equations defined over the interval 0 through 10.

4. Solve this problem from the Rhind papyrus using the method of false position: A quantity and its 1/7 added together become 19. What is the quantity?

5. Write 1 million using the base-60 cuneiform writing of the ancient Babylonians.

6. Positions in a 4 × 4 × 4 three-dimensional tic-tac-toe game are located using the number triples (x, y, z) with the variables taking on the values 1, 2, 3, 4. Find the seven winning 4-position arrangements that contain (2, 2, 2).

7. Find the tenth term in the Fibonacci sequence that starts with 3 and 7. Is the sum of the first 10 terms the same as 11 times the seventh term?

8. Verify the rule of double false position. Assume the two guesses x_1 and x_2, when substituted into the linear equation $ax + b = 0$, give results of r_1 and r_2 and show that $x = (r_1x_2 - r_2x_1)/(r_1 - r_2)$.

9. Start with a four-digit number with all digits different. Prove algebraically that the sum of all possible three-digit numbers that can be formed from these digits divided by the sum of the digits of the original number must equal 666.

10. Derive a method that makes use of successive differences to find the function $f(x) = ax^2 + bx + c$ given $f(1)$, $f(2)$, and $f(3)$. Then apply the method for the quadratic $f(x) = 3x^2 - 8x + 5$.

11. Use your calculator to evaluate e using the first six terms given in this series.

$$1 + \frac{1}{1} + \frac{1}{2 \cdot 1} + \frac{1}{3 \cdot 2 \cdot 1} + \frac{1}{4 \cdot 3 \cdot 2 \cdot 1} + \frac{1}{5 \cdot 4 \cdot 3 \cdot 2 \cdot 1} + \cdots$$

12. Use your calculator to evaluate $\sqrt{2}$ using the continued fraction

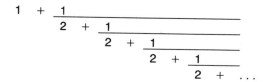

$$1 + \cfrac{1}{2 + \cfrac{1}{2 + \cfrac{1}{2 + \cfrac{1}{2 + \ldots}}}}$$

ACTIVITIES

1. Describe two different formats that can be used in teaching the six partial products that are formed when multiplying the binomial $a + b$ and the trinomial $c + d + e$.

2. Illustrate with an appropriate drawing how the trinomial $3x^2 + 7x + 2$ can be factored geometrically.

3. Start with a large equilateral triangle cut from paper. Fold each vertex to the midpoint of the opposite side. Repeat the process with each new equilateral triangle formed. If it could be folded, what fractional part of the original would be represented by the tenth such triangle formed?

4. Plan a short lesson on squaring the trinomial $a + b + c$ that makes use of a geometric model cut from a paper square.

5. Discuss the advantages in having an aid to demonstrate the algebraic concept of factoring.

6. Describe how the game of checkers might be used to introduce coordinate geometry into a mathematics class.

7. Develop a lesson centered around translating a transparency overlay of the parabola $y = x^2$ on a grid base.

8. Construct a transparency that can be used to illustrate the family of lines with $(0, -6)$ as their y-intercepts.

9. Show how the equation of a straight line in slope-intercept form can be reviewed using the overhead projector and an overlay on a transparency grid.

10. Construct a transparency similar to the one described on page 185 for drill and practice in multiplying binomial factors.

11. Construct a model showing all complex points, both real and imaginary, in $y = x^2 + 2$ using the method shown on page 165. Be sure to locate and identify the two complex roots.

12. Prepare a report on the apparent relationship between the Fibonacci sequence and certain aspects of nature.

13. Construct a teaching aid other than those described in this chapter that can be used in the teaching of algebra.

14. Run the BASIC program on page 187 for Leibniz's series for pi using input values of 100, 1000, and 10,000.

15. Identify a series expansion for pi that converges more rapidly than Leibniz's series expansion.

READINGS AND REFERENCES

1. Read Chapter 5 on the history of algebra in *Historical Topics for the Mathematics Classroom*, the 31st Yearbook of the National Council of Teachers of Mathematics, 1969. Prepare a report on three of the topics included in the chapter.

2. Read Chapter 3 in *Mathematics and the Imagination* by Edward Kasner and James Newman, New York: Simon & Schuster, 1967. Summarize the key information given on the numbers π, i, and e that would be suitable for use in teaching an upper level mathematics course in the secondary school.

3. Read Chapter 2 in *What Is Mathematics?* by Richard Courant and Herbert Robbins, New York: Oxford University Press, 1967. Outline the important topics covered in their section on complex numbers.

4. A major source of information on the mathematics of the ancient Egyptians is the Rhind Papyrus. Problem 48 in the collected set of arithmetical problems in the third part of the papyrus compares the area of a circle with that of the circumscribed square. Read a translation of the problem, describe the procedure, and the resulting approximation used for pi. An excellent reference is *The Rhind Mathematical Papyrus* by A. B. Chace recently reprinted by the National Council of Teachers of Mathematics in 1979.

5. Read about the mathematical achievements of the sixteenth century in a history of mathematics book. Summarize the key accomplishments in arithmetic and algebra. One established reference in the field is *An Introduction to the History of Mathematics* by Howard Eves, New York: Holt, Rinehart, and Winston, Inc., 1969.

6. Read Chapter 3 in *Men of Mathematics* by E. T. Bell, New York: Simon & Schuster, 1965. Prepare a report on the mathematician described in this chapter entitled "Gentleman, Soldier, and Mathematician."

7. It is important that mathematics teachers share with their students the fact that mathematicians emerge from many different nationalities and backgrounds with many varying abilities and insights. Perhaps one of the most remarkable stories is that of Ramanujan. Read the commentary on this mathematician from India in Volume I of *The World of Mathematics* by James Newman, New York: Simon & Schuster, 1956.

8. The *Mathematics Teacher*, a publication of the National Council of Teachers of Mathematics, contains many interesting applications of the microcomputer to the teaching of algebra. Read two such articles, run the corresponding programs, and evaluate their potential effectiveness.

9. Review *Computer Programming in the BASIC Language*, Second Edition, by Neal Golden, Orlando, Florida: Harcourt Brace Jovanovich, 1981. Do two of the Round Exercises given at the end of each of the first four chapters categorized under the heading "Algebra One."

Classroom Aids and Activities: Geometry

Chapter 7

Geometry is a subject rich in motivational material that can capture the attention and imagination of students from the earliest grades on through the advanced secondary level and beyond. Activities in informal geometry in the middle and junior high school can be used to introduce new ideas and reinforce old ones. Theorems from high school geometry can have their beginnings with concrete, manipulative experiences that give meaningful insight and understanding before structured proof. Visualization activities can help stretch students' minds and make them more flexible and creative. Equally important, geometric thinking and analysis give students powerful problem-solving tools, often offering a fresh new outlook to a challenging situation.

This chapter begins with some short motivational activities for geometry and then examines material related to the teaching of polygons, polyhedrons, and the conic sections. Visualization, measurement, and construction activities and experiments follow along with some ideas for the overhead projector plus calculator and computer applications.

7.1 MOTIVATING GEOMETRY

This section lists some strategies for teaching mathematics and then illustrates them with simple classroom experiences in geometry.

Encourage Students to See Things from a Geometric Vantage Point

This classification of digits with the letters A, B, and C focuses on one of their properties that students may not have noticed before. How would you classify the digit 9?

A	1	4	7
B	2	5	
C	0	3	6	8

Use Geometry as a Vehicle for Problem-Solving Activities

Have every student divide a 1-inch x 4-inch strip of paper into quarters as shown, labeling the four parts on both sides as 1, 2, 3, and 4. Imagine all the different ways the strip can be folded into a stack of squares. Imagine all the four-digit numbers possible using the digits 1, 2, 3, and 4. Which can be formed from the stack of squares, reading the digits from the top down? Which cannot be formed?

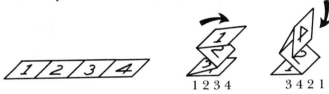

1 2 3 4   3 4 2 1

Illustrate Applications of Geometric Properties

It's one thing to know the converse of the Pythagorean theorem, but it's quite another thing to show it in action. Take a rope and put knots in it every foot. Then see if your students can figure out how to use it to form a right angle. The ancient Egyptians used this very method in marking off the right-angled corners of their fields each year after the annual flooding of the Nile River. All it takes is knowledge of a 3–4–5 right triangle.

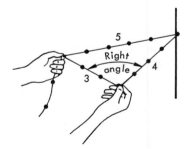

Use Models for Illustration and Comparison

When teaching about cylinders, take an 81/2-inch x 11-inch sheet of paper. Roll it end-to-end lengthwise and then widthwise, as shown. Which cylinder has the greater volume? Estimate what percent more. Does the smaller have the same percent less?

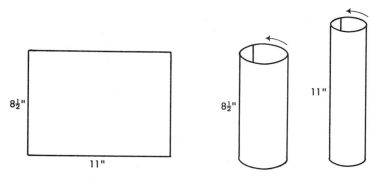

Put Things in Your Students' Hands

Consider metric units of volume. If students can see, touch, and manipulate models of a cubic meter, centimeter, and millimeter, they are far more likely to remember them. Students are always impressed with the size of a model of a cubic meter (m^3). Even a cubic centimeter (cm^3) carries little real meaning unless students get a visual impression of it by holding some centimeter cubes in their hands. After you pass some out, let the class see how many different solids can be formed by joining centimeter cubes together face-to-face. Use 2 cubes, 3 cubes, and then 4 cubes. Watch for mirror images that are identical and look for those that are not.

To dramatize just how small a cubic millimeter (mm^3) is, cut some from the square ends of flat toothpicks. Use a razor blade. You can get two or three good models off each toothpick. Drop one of them in each student's hand. Don't tell them where they came from. Just let them react to their size.

Get the Most You Can Out of Your Classroom Demonstrations

In doing a classroom demonstration, we often don't stay with it long enough to realize the full results. Here is an example.

Fold a square piece of paper in half vertically as shown. Before cutting the result in half, ask how many pieces you will get. If your students say 2, cut vertically. If they say 3, cut horizontally. Students need these surprises once in a while.

Fold in half	Two pieces	Three pieces

The original visualization problem is done. But don't stop.

Suppose the cut is made vertically down the middle. Three rectangles are formed, two small and one large. Without first showing the pieces, ask how their areas compare. Follow that easy question with one about how their perimeters compare. It's a nice little arithmetic or algebra exercise to show that the answer is 5 to 6, smaller to larger.

Now stretch the activity into the problem-solving stage with this question.

How can the folded square be cut vertically such that the ratio of perimeters of the rectangles, smaller to larger, is 1 to 2?

7.2 POLYGONS

The classification of polygons is one of the major objectives in geometry for the elementary and junior high school. Once definitions have been given, students need to see polygons in a wide variety of experiences so that they can recognize them and become familiar with their properties and applications. These activities can take the form of motivational puzzles, basic concept review and extension, and challenging problem-solving experiments. Several illustrations are given here that involve hands-on materials for student use.

Tangrams

The tangram puzzle given on page 96 involves certain problem-solving skills such as trial and error. It also requires visualization skills since the final polygons cannot be seen physically until the pieces are properly arranged. If more visualization practice is sought, but at perhaps a simpler level, use this activity with the original lettered square before the pieces are cut out. It requires recognizing existing polygons but not creating new ones.

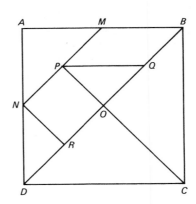

Use vertices to name all the convex polygons in the lettered square.

Triangles	BCO, CDO, AMN, DNR, OPQ, ABD, BCD
Quadrilaterals	
squares	ABCD, NPOR
parallelograms	BQPM
trapezoids	BOPM, BRNM, QRNP, ODNP, BDNM, PQDN
others	BCPM, CDNP, ABRN
Pentagons	BCDNM

Review congruency by noting that quadrilaterals BCPM, CDNP, and ABRN are congruent. See what other sets of congruent polygons in the figure can be found by your students.

Take advantage of geometric activities such as these to review arithmetic skills as well. Here are some examples relating tangrams to percents.

What percent of the original square is represented by the area of each of these polygons?

triangle BCO	25%
square NPOR	12 1/2%
quadrilateral BCPM	43 3/4%

As a helpful hint, think of adding the necessary lines to divide the entire tangram square into 16 small triangles congruent to DNR and OPQ. Each such triangle is 1/16 of the original square.

Some interesting perimeter questions can be asked as well, once students have studied radicals and the Pythagorean theorem.

Assuming the original square is a unit square, find the perimeter of each of these polygons.

triangle BCO	$1 + \sqrt{2}$
square NPOR	$\sqrt{2}$
quadrilateral BCPM	$1\ 1/2 + \sqrt{2}$

Of course, the perimeter can also be found for every polygon formed by rearranging the original 7 pieces of the tangram puzzle.

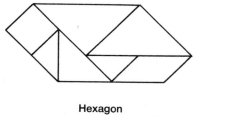

Hexagon $2 + 2\sqrt{2}$

Other measurement activities can be found in section 7.7.

Congruency

Many believe congruency is an easy concept, quickly mastered. In general, this is true but the orientation of the figures can have much to do with successful recognition. This activity is designed to reinforce this aspect of the concept of congruency.

Cut some 4 x 4 squares from graph paper. Then find how many different ways you can cut them in half along the grid lines. Six different polygonal shapes are possible from the halves.

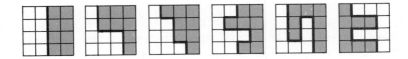

Students can explore the same problem by drawing on graph paper. However, the advantage to cutting out the pieces is that they can physically reorient the pieces and immediately check for congruence by placing one on top of the other to see if they coincide.

Several interesting extensions are possible. The two halves, of course, have the same areas. But what about their perimeters? Compute them starting with the 4 x 4 square. Students may be surprised to find that two pairs of halves have perimeters greater than that of the original whole square!

Classifying Triangles

Start with 6 toothpicks placed on the overhead projector. Have students come up and arrange any number of them, end-to-end, to form triangles. Identify each different solution as a number triple corresponding to the number of toothpicks on each side. Two equilateral triangles (1,1,1 and 2,2,2) are possible plus an isosceles triangle (1,2,2). Don't count (2,1,2) and (2,2,1) separately as they give the same triangle as (1,2,2). Watch for those who incorrectly try to offer another isosceles triangle (1,1,2) and a scalene triangle (1,2,3).

Now ask how many more toothpicks need to be added so that a scalene triangle is possible. Add them to those already on the overhead projector and see if they can find and classify all eight triangles possible using some or all of the toothpicks available.

Finally, turn off the overhead projector. Have each student make a list, using numbers, of all possible triangles that can be formed with any number up to 12 toothpicks. Be sure they classify each triangle as equilateral, isosceles, or scalene.

EQUILATERAL TRIANGLES	ISOSCELES TRIANGLES	SCALENE TRIANGLES
1,1,1	1,2,2	2,3,4
2,2,2	1,3,3	2,4,5
3,3,3	1,4,4	3,4,5
4,4,4	1,5,5	
	2,3,3	
	2,4,4	
	2,5,5	
	3,4,4	

Geoboards

The geoboard can serve as an excellent aid for individual exploration and experimentation. It is so easy to use that students can become actively involved at a concrete level in the creative, imaginative aspects of geometric discovery. Classroom sets of geoboards are readily available commercially or they can be constructed by the school's shop classes with little difficulty.

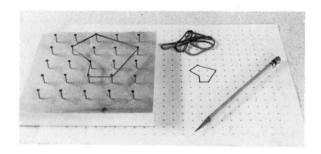

Dot paper and grid paper are useful supplements for the geoboard and should be readily available to students so that they can easily record specific figures when needed. A geoboard made of clear plastic can serve as a very effective and dynamic aid for the teacher when used with an overhead projector.

Many geometric concepts of the early elementary grades through the secondary years can be developed through geoboard activities and experiments. Some applications are suggested here in the format of student worksheets. The first deals with squares of various sizes and the second with congruency. The third offers a discovery experience leading to an interesting algebraic formula relating the area A of a polygon to the number of points on the boundary b and in the interior i. It is called *Pick's Formula*.

$$A = b/2 + i - 1$$

Squares

Use your geoboard.

1. Form a square using the segment shown as a side.

2. Form another square using this segment as a side.

3. Find all the squares of different sizes that you can form on the geoboard. In all there are eight possibilities.

4. Draw the eight squares here, arranging them in order, smallest to largest.

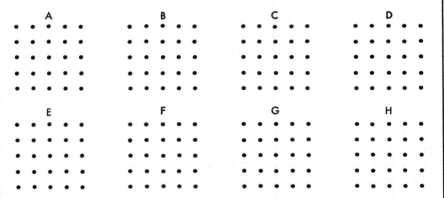

5. Using the total area of the 4 x 4 geoboard as 16 square units, give the area of each square drawn in question 4.

    A＿＿   B＿＿   C＿＿   D＿＿   E＿＿   F＿＿   G＿＿   H＿＿

6. Now find the perimeter of each square using 1 unit as the distance between adjacent points on the geoboard.

    A＿＿   B＿＿   C＿＿   D＿＿   E＿＿   F＿＿   G＿＿   H＿＿

Congruency

1. Form all the squares congruent to this one that you can. How many can you find?

2. How many rectangles can you find that are congruent to this one?

3. How many right triangles can you form that are congruent to this one?

4. Form on your geoboard a polygon congruent to this one but translated one unit left.

5. Form a polygon congruent to the one above but flipped about the central horizontal axis.

6. Form a polygon congruent to the one above but rotated 90° clockwise.

7. How many different positions are possible on the geoboard for the pentagon shown in question 4? Count all translations, flips, and rotations but don't count duplications of positions.

Pick's Formula

Copy this table on a piece of paper. Construct each polygon described using the geoboard. Then enter the required values for *b* and *i* for that polygon in the table.

Let *b* be the number of points on the boundary of the polygon.
Let *i* be the number of points in its interior.

POLYGON	*b*	*i*	AREA
1			
2			
3			
4			
5			
6			
7			
8			

1. A 1 x 1 square
2. A 1 x 3 rectangle

3. A right triangle with legs 1 and 4
4. A right triangle with legs 1 and 3

Study the entries in the table thus far. Can you discover how to get *A* from *b* and *i*?

5. A 2 x 2 square
6. A 3 x 4 rectangle

7. A right triangle with legs 2 and 3
8. An isosceles triangle with two equal sides of 4

Study the entries in the table now. Can you discover how to get *A* from *b* and *i*?

9. Is *b* always greater than *A*?
10. Is *b*/2 always greater than *A*?
11. Is *b*/2 + 1 always greater than *A*?
12. How is *A* related to *b* and *i*?

$$A = \underline{\hspace{3cm}}$$

Test your final answer using another figure of your own choice.

7.3 POLYHEDRONS

Students need to know the definitions of prisms and pyramids along with their properties and formulas. They also need problem-solving experiences and applications involving these and other solids formed with flat faces bounded by polygons. This section covers some activities with various polyhedrons while the next section covers some methods of constructing appropriate models.

Sketching

Models play an important role in the study of polyhedrons but students also need to be able to "see" these 3-dimensional solids and their properties from 2-dimensional representations. With that in mind, take time to share with your class some procedures for sketching prisms and pyramids. One of the most useful makes use of graph paper.

By copying figures drawn on a grid, students can quickly and accurately sketch their own while preserving parallelism and aspects of congruency. Prisms have congruent bases and parallel and congruent lateral edges and these are clearly evident in the sketch. Have your students copy these figures first and then create some others of their own.

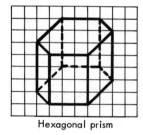
Hexagonal prism

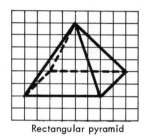
Rectangular pyramid

Prisms and Pyramids

Some students can look at 2- and 3-dimensional representations of prisms and pyramids and clearly see the separate faces that make up the solid. Others cannot. For them, this particular activity would be especially valuable.

Cut the indicated number of pieces for each polygon from construction paper of a different color.

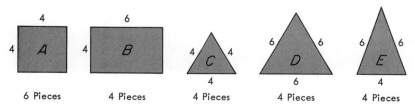

4	6		6 6	6 6
4 _A_	4 _B_ 4	4 _C_ 4	_D_	_E_
		4	6	4
6 Pieces	4 Pieces	4 Pieces	4 Pieces	4 Pieces

Divide the class into two teams. Take turns having a member of a team come up and demonstrate how particular prisms and pyramids can be formed from a selected set of pieces. Use the appropriate letters to keep record of those formed. Keep the score for each team. Require proper identification of each solid formed. Allow for challenges. Encourage speed.

A surprising number of possibilities exist, some easier to find than others. All can be readily verified by assembling the pieces correctly.

RECTANGULAR PRISMS	RECTANGULAR PYRAMIDS
AAAAAA	ACCCC
AABBBB	AEEEE
	BDDEE
	BCEEE

TRIANGULAR PRISMS	TRIANGULAR PYRAMIDS
AAACC	CCCC
ABBEE	DDDD
BBBCC	CEEE
BBBDD	DDEE
	EEEE

For a simpler version, delete the D pieces (6-6-6 equilateral triangles). For a more difficult version, add some new F pieces (4-4-6 triangles).

Euler's Formula

Although known to Descartes more than 100 years earlier, this simple relationship among the number of vertices, faces, and edges of a polyhedron was independently discovered by Euler and now bears his name.

**EULER'S
FORMULA**

> If V is the number of vertices, E the number of edges,
> and F the number of faces of a simple polyhedron, then
> $V + F = E + 2.$

This experiment offers students an opportunity to explore these numbers, V, F, and E for various polyhedrons. Hopefully, it will lead them to discover and support Euler's formula.

A polyhedron is a figure in space that has flat faces bounded by polygons. The prisms and pyramids shown here are examples of polyhedrons.

1. Count the number of vertices, faces, and edges for each prism and pyramid shown. Then record the numbers in the table.

POLYHEDRON	NUMBER OF		
	VERTICES	FACES	EDGES
cube			
tetrahedron			
pentagonal prism			
square pyramid			
hexagonal prism			

2. Study the values in the table. For each of these polyhedrons is the number of edges E greater than the number of vertices V and greater than the number of faces F? Is E always less than the sum of V and F? Can you discover a relationship among the number of edges E and the number of vertices and faces V and F? Try to express this relationship in symbols.

3. Support your formula from step 2 using the V, F, and E from each of these polyhedrons.

4. Now imagine a small piece cut off each of the eight corners of a cube. Mentally count the number of vertices, faces, and edges of the resulting polyhedron. Do V, F, and E support Euler's formula?

5. Consider a prism with n-gons as bases. Express V, F, and E in terms of n and see if you can support Euler's formula algebraically.

6. Try the same with a pyramid with an n-gon as its base.

Regular Polyhedrons

The literature in the history of mathematics is rich with stories about regular polyhedrons. Their unique special beauty and symmetry has intrigued men in all ages of time. Some of the solids were probably known to the ancient Egyptians, and the Pythagoreans (ca. 500 B.C.) discovered others. They were later known as the Platonic solids when the Greek, Plato (ca. 400 B.C.), wrote of their role in the design of matter with this mystical association to the four so-called "elements" of the universe:

tetrahedron	fire
hexahedron (cube)	earth
octahedron	air
icosahedron	water
dodecahedron	universe

Euclid (ca. 300 B.C.) devoted the last of the 13 books in his famous *Elements* almost entirely to these solids. He included a proof that only five such solids exist.

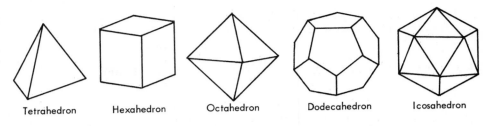

Tetrahedron   Hexahedron   Octahedron   Dodecahedron   Icosahedron

Not many students, looking at solids such as these, are aware of their various parts. Ask the students to count vertices, faces, and edges. See who can come up with simple counting techniques.

Edges are joined at vertices and faces joined at edges. A dodecahedron has 12 pentagonal faces. The separate edges of the individual faces (12 × 5 = 60) are paired to form the edges (60/2 = 30) of the solid. The separate vertices (12 × 5 = 60) are joined three at a time for the (60/3 = 20) vertices of the solid.

REGULAR POLYHEDRON	SHAPE OF FACES	NUMBER OF		
		VERTICES	FACES	EDGES
Tetrahedron	equilateral triangles	4	4	6
Hexahedron	squares	8	6	12
Octahedron	equilateral triangles	6	8	12
Dodecahedron	regular pentagons	20	12	30
Icosahedron	equilateral triangles	12	20	30

Each of these number triples satisfies Euler's formula, $V + F = E + 2$.

An interesting and challenging question related to the regular polyhedrons involves coloring. Faces of each solid are painted individually, each with a single color. What is the least number of colors needed to paint each solid such that no two adjacent faces have the same color? The answers are surprising!

This semiregular polyhedron has 6 squares and 8 equilateral triangles as faces, with two of each alternating at each of its 12 vertices. How many edges does it have?

7.4 MODELS

Students enjoy making their own models from the simplest prism or pyramid to the most complex semiregular or stellated polyhedron. This section illustrates a variety of possible techniques, some more suitable at one level with one kind of figure than at another with a different figure.

Paper Strips

Cut some paper strips 6 cm wide and 12 cm long. Fold them in different ways to form simple models of a variety of prisms.

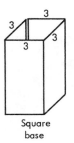

Square base

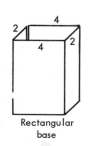

Rectangular base

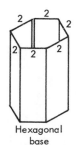

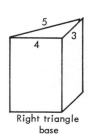

Hexagonal base

Right triangle base

Toothpicks

Glue toothpicks together to form the edges of some pyramids.

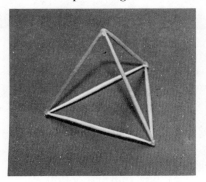

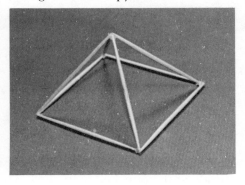

Straws and Twist-ties

Cut some straws to the appropriate lengths to serve as edges. Twist some twist-ties together to join the straws at the vertices. Three twist-ties twisted together are needed at the apex of the pentagonal pyramid shown. Two are needed at each of the other vertices.

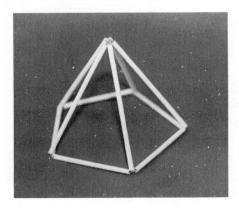

Folding Paper Hexagons

Cut some regular hexagons from heavy paper. Fold each on its three main diagonals and cut along one to the center. As you curl up the hexagon, models of a pentagonal, square, and triangular pyramid are formed. A paper clip can be used to hold the pyramid in place.

 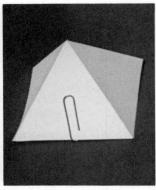 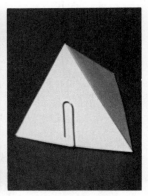

Use the three pyramids formed to help the students discover and generalize the relationship between the number of vertices in the base and the total number of vertices and of edges. A pyramid with an n-gon as a base has $n + 1$ vertices and $2n$ edges.

File Cards and Rubber Bands

This method can be used to carefully assemble more involved models. Draw each face on a file card. Punch holes at each vertex. Cut along the outside of the punched holes. Fold back at the edges marked. Then attach adjacent faces by joining corresponding edges with rubber bands.

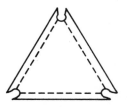

Cut on the solid lines and fold on the dashed lines. Make four separate triangular faces. Attach them with rubber bands to form the regular tetrahedron shown.

Another similar construction technique uses equilateral triangles folded from circular pieces of paper or cardboard. Fold the tabs out and staple them together.

Nets

More complicated figures are often formed by first constructing a net showing the layout of adjacent faces from oaktag or cardboard. Cut around the boundary and score along the interior edges to facilitate folding. Then assemble and tape or glue using flaps.

Here are nets for the five regular polyhedrons. One way to copy them is to cut a template in the shape of each of the basic faces from a file card. The net can then be laid out accurately by marking each vertex of each face with a pencil or pin.

One set of possible patterns for construction is shown here.

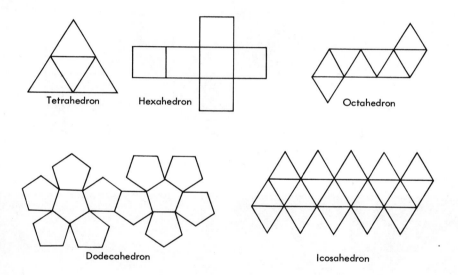

| Tetrahedron | Hexahedron | Octahedron |

| Dodecahedron | Icosahedron |

The skills involved in the careful and accurate construction of these solids are often neglected in the mathematics class. As an activity, it can be justified for this reason alone. However, work with these solids can help develop still another important, often neglected, skill, that of three-dimensional visualization.

Plaiting

To make most models of polyhedrons rigid, the edges must be taped or extra tabs must be glued together. However, this process allows assembling the cube without glue or tape. The technique is known as plaiting. With proper instruction, students should be able to assemble their own models.

Fold strip *A* over strip *B* and under strip *C*, positioning faces 1 and 5 of the cube. Tuck strip *B* over strip *C* and under strip *A*, positioning faces 4 and 6 of the cube. Fold strip *C* over flap *A* and under strip *B* positioning faces 2 and 3. Now tuck flap *B* into the slot under face 1. The cube with faces numbered 1 through 6 should be assembled.

Folding a Paper Circle

Teachers often associate a model with a particular grade-level topic. However, most geometric models have a wide variety of applications from the simple to the challenging, elementary through secondary. Consider this folding of a paper circle into a model of a truncated tetrahedron.

1. Start with a 12-inch circle cut from a sheet of newspaper.

2. Fold two diameters to locate the center of the circle.

3. Take the edge of the circle and fold it back onto the center three times in such a way as to form an equilateral triangle.

4. Fold one vertex to the midpoint of the opposite side for an isosceles trapezoid, two for a rhombus, and all three for a small equilateral triangle.

5. Open the model up so that the four equilateral triangles form the faces of a regular tetrahedron.

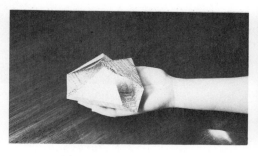

6. When you fold each vertex of the large triangle into the center, a regular hexagon is formed.

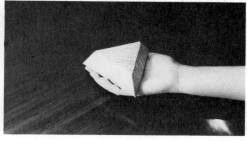

7. Now rest the folded paper loosely in the palm of your hand.

8. Tuck the three upper flaps together to form a model of a truncated tetrahedron.

When the process is demonstrated by the teacher, students at most levels can follow the steps and construct their own model. But where and how can it be used in the classroom? The following possibilities illustrate just how widely applicable this activity can be.

1. Classifying polygons   By the time step 6 is reached in the construction, these different polygons have been formed:

equilateral triangle	isosceles trapezoid
rhombus	regular hexagon

By refolding along various existing folds, quite a collection of polygons can be formed. Supply triangular grid paper as shown and have students copy each polygon they find. The complete set of 10 is shown here.

By changing the process just slightly to allow folds under as well as over, a whole new set of concave polygons is added to those convex ones given above. Can you find all 10 concave polygons?

2. Fractions   Call the area of the original equilateral triangle 1. Then the isosceles trapezoid has an area of 3/4, the rhombus 1/2, and the small equilateral triangle 1/4. Can you show by the way the hexagon was folded that it must have an area of 2/3?

Every one of the 20 polygons mentioned above has a corresponding fraction that can be used to describe its area. The area of the pentagon shown here can be found by subtracting 1/4 and 1/9 from 1.

$$1 - 1/4 - 1/9 = 23/36$$

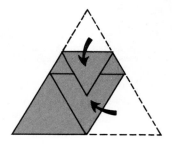

3. *Constructions* Compare the paper-folding construction with the corresponding compass and straightedge construction at various stages.

How can you locate the center of a circle?
How can you locate the midpoint of a side?
How can you locate an altitude of an equilateral triangle?

4. *Theorems* State the geometry theorem or theorems illustrated at these steps:

Step 3, where the circle is folded in three times to form an equilateral triangle
Step 4, where four congruent triangles are formed when the midpoints of the sides are joined
Step 5, where the center of the base is 2/3 the distance from each vertex to the opposite edge
Step 6, where each small equilateral triangle has an altitude 1/3 that of the original large one
Step 8, where the height of the tetrahedron cut off the top is 2/3 and the volume 8/27 that of the original tetrahedron

5. *Algebra* If the original circle has a diameter of 12 inches, find each of these lengths:

An altitude of the original equilateral triangle
A side of the original equilateral triangle
The height of the original tetrahedron
The height of the truncated tetrahedron

6. *Advanced algebra* What are the surface area and volume of the truncated tetrahedron? Careful analysis of the problem can lead to an interesting plan of attack.

If you study the last two steps of the construction, it is apparent that the area of the truncated tetrahedron is exactly 7/9 that of the original tetrahedron, which is nothing more than the area of the original equilateral triangle.

When considering the volume of the original large tetrahedron, note that the height of the small tetrahedron cut off the top is 2/3 that of the large tetrahedron. Since the two tetrahedrons are similar in shape, it follows that the ratio of their volumes must be $(2/3)^3$ or 8 to 27. The remaining portion must therefore be 19/27 of the original volume.

The area and volume of the original tetrahedron can also be found more directly using the fact that the diameter of the initial circle is 12 inches. However, the computation is more tedious.

A Folding Puzzle

Here is a puzzle for students who enjoy this type of activity. Show how an 8-inch strip of paper 1 inch wide can be folded to form a cube. For a somewhat harder version, mark eight ×'s on one side of the strip and fold a cube with all faces showing ×'s.

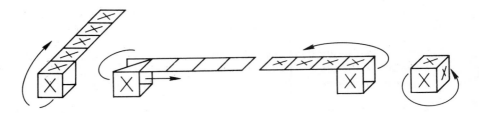

7.5 CONIC SECTIONS

One of the most important sets of curves in mathematics is the conics. The ancient Greek Apollonius (225 B.C.) wrote a treatise entitled *Conic Sections* in which he described how the ellipse, parabola, and hyperbola can be formed by passing a plane through a cone at various angles. Later, with the introduction of coordinate geometry, the conic sections were expressed algebraically as well. Several of the many methods for constructing these curves are given here.

Most mathematics teachers are familiar with the wooden models of cones that can be taken apart to show cuts in the shape of circles, ellipses, parabolas, and hyperbolas.

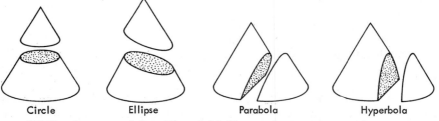

| Circle | Ellipse | Parabola | Hyperbola |

While these models offer vivid illustrations, they are by no means unique since many examples of each curve can be found on the very same cone. Every plane that cuts a cone forms a conic section.

Start with a cutting plane through some point P on the cone. If it is perpendicular to the axis, a circle is formed. At the exact position where it is parallel to an element of the cone, a parabola occurs. All other positions produce ellipses or hyperbolas.

These positions are illustrated in this head-on cross-section view.

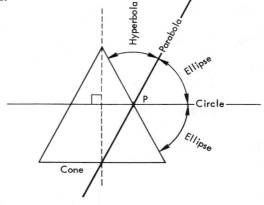

A simple flashlight can be a useful aid when teaching conics. Held at different angles, the circular reflector can project on the wall a circle, ellipse, parabola, and hyperbola. As you move it through the different angles, you can see the curves change much as illustrated above with the cone.

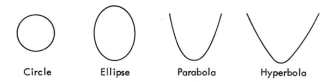

Circle   Ellipse   Parabola   Hyperbola

A construction for an ellipse using a string loop was shown in Chapter 2 on page 48. If the string loop is 18 inches long and the tacks are 8 inches apart, the distance from any point on the curve to the two tacks, which are the foci, is fixed at 10 inches. The semimajor axis is 5, the semiminor axis is 3, and the equation of the ellipse is

$$\frac{x^2}{25} + \frac{y^2}{9} = 1$$

As the tacks get closer together, the curve, while remaining elliptical, approaches a circle with radius 9 inches. Only when the two points coincide will a circle be formed.

To illustrate this transition in reverse, consider a plane slicing a can. A cut perpendicular to the axis of the can forms a circle with its center at the axis. But tip the cutting plane and an ellipse is formed with the axis located midway between the two foci.

Try making a model from an empty frozen juice can cut with a band saw in the school shop.

Although most algebra courses include some treatment of the conics from an analytic approach, little is usually done with them at a more informal level. The following experiments illustrate some simple methods for constructing the conics. They offer interesting and valuable activities for students at both the junior and senior high levels.

EXPERIMENT 1   Conics from Circles and Lines

Material

A set of worksheets with sets of concentric circles and parallel lines as shown.

Directions

1. PARABOLA: Locate and mark the intersection of circle 1 and line 1. Next mark the two intersections of circle 2 and line 2. Then repeat for circle 3 and line 3, circle 4 and line 4, and so on. Connect these points with a smooth curve, a parabola.

Take any point on the curve. Is it the same distance from line 0 and the center of the circles?

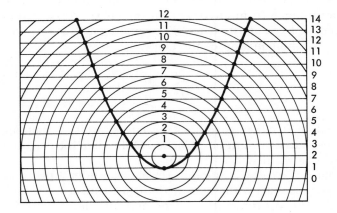

2. ELLIPSE: Locate and mark the points where circles numbered 4 and 8 intersect. Each of these four points is 4 units from one center and 8 units from the other. The sum of the distances from the two centers is 4 + 8, or 12. Next mark the four intersection points of circles 5 and 7. They, too, are located such that the sum of their distances from the two centers is 12. Now mark all other sets of points where the sum of the distances from the two centers is 12. For example, 3 and 9 and 2 and 10. Connect the points with a smooth curve, an ellipse.

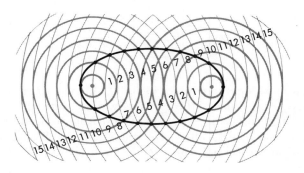

3. HYPERBOLA: Locate and mark points in a similar fashion, but this time with those that always have a difference of 6. For example, 8 and 2, 9 and 3, 10 and 4, and so on. The two smooth curves formed by connecting these sets of points are branches of a hyperbola.

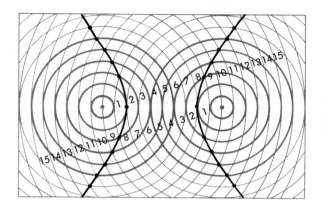

Analysis

These constructions yield the desired conics because they are based on the following loci definitions:

Parabola—the set of points such that each is equidistant from a fixed point and a fixed line.

Ellipse—the set of points such that the sum of the distances of each from two fixed points is constant.

Hyperbola—the set of points such that the differences of the distances of each from two fixed points is constant.

Try drawing a family of conics on each worksheet. Just measure from a different line than zero for more parabolas. Let the sum and difference be some values other than 12 and 6 for more ellipses and hyperbolas.

EXPERIMENT 2   Paper Folding the Conics

Material

A sheet of paper with a line on it and two sheets with circles.

Directions

On a sheet of paper mark a point as shown. Repeatedly fold so the point coincides with the line or circle. The more creases made, the more apparent the conic becomes and the smoother the curve.

Fold various points of the line onto a point not on the line. The creases form tangents to a parabola.

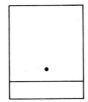
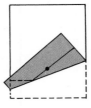

Parabola

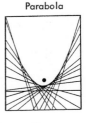

Fold various points of the circle onto a point inside the circle. The creases form tangents to an ellipse.

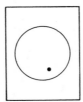
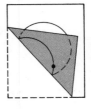

Ellipse

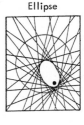

Fold a point outside the circle onto various points of the circle. The creases form tangents to a hyperbola.

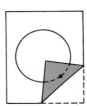

Hyperbola

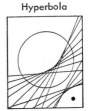

Excellent examples of these curves can be formed using waxed paper where the creases are more vivid. These can then be easily projected on the overhead projector.

Analysis

Geometry students should be able to prove that these constructions give the conics.

1. PARABOLA. Point P is folded onto the line at P'. The crease is tangent to the parabola at point A. The focal point of the parabola is P.

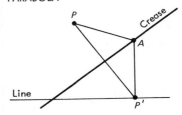

PARABOLA

The crease is the perpendicular bisector of segment PP', so PA and $P'A$ are equal in length. Since $P'A$ is perpendicular to the given line, it measures the distance from A to that line. Therefore, point A is equidistant from point P and the line and must lie on the parabola.

2. ELLIPSE. Point *P* is folded onto circle *O* at *P'*. The crease is tangent to the ellipse at point *A*. The foci of the ellipse are at *P* and *O*.

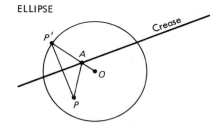

ELLIPSE

The crease is the perpendicular bisector of segment *PP'*, so *PA* and *P'A* are equal in length. Hence *PA* plus *AO* are equal in length to *P'A* plus *AO*, which is the radius. Since the radius is constant in length, the sum of the distances of *A* from *P* and *O* is constant. Thus point *A* must be on the ellipse.

3. HYPERBOLA. Point *P* is folded onto circle *O* at *P'*. The crease is tangent to the hyperbola at *A*. The foci of the hyperbola are at *P* and *O*.

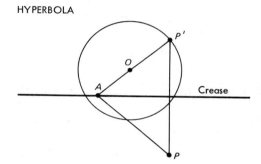

HYPERBOLA

The crease is the perpendicular bisector of segment *PP'*, so *PA* and *P'A* are equal in length. Hence *PA* less *AO* is equal in length to *P'A* less *AO*, or simply *OP'*. But *OP'* is a radius and constant in length. Therefore, the difference of the distances of *A* from *P* and *O* is constant. Thus point *A* must be on the hyperbola.

7.6 VISUALIZATION ACTIVITIES

One valuable problem-solving skill is the ability to think geometrically—to visualize a geometric interpretation or result mentally without having the physical model to hold, position, or otherwise analyze. The activities in this section are directed toward that goal. They may still involve models, but these serve now as a method of check or support following a conjecture arrived at through mental visualization.

Folding Paper Squares

This activity is excellent for both classroom demonstration and individual student involvement. It calls for mental observations that at the same time review basic geometric definitions and numerical relationships which, when needed, can immediately be reinforced with the physical model.

Imagine a paper square with vertices labeled in order as *A, B, C,* and *D*. Visualize the figure that will result from each of these folds.

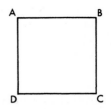

Single Folds

1. Fold vertex *A* to vertex *B*.
2. Fold vertex *A* to vertex *C*.
3. Fold vertex *A* to the center of the square.
4. Fold vertex *A* to the midpoint of side *AD*.

Successive Folds

5. Fold *A* to *B* and then *B* to *C*.
6. Fold *A* to *C* and then *B* to *D*.
7. Fold *A* to *C* and then *B* to *C*.
8. Fold *A* and *C* to the center of the square.
9. Fold *A* and *B* to the midpoint of side *AB*.
10. Fold *A* to the midpoint of side *AB* and then *B* to the midpoint of side *BC*.

As a follow-up, tie these same visualization questions to arithmetic and algebra.

> If the original square has an area of 1, what fraction is represented by each result?
>
> If the original square measures *n* x *n*, what algebraic expressions represent the area and perimeter of each result?

At a more challenging level, use a regular hexagon or regular pentagon.

Fold a pair of opposite vertices of a regular hexagon to the center. What figure results? What fractional part of the area of the original hexagon remains?

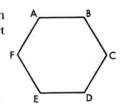

In this regular pentagon fold *A* to *C* and then fold *B* to *D*. What percent of the area of the original pentagon remains?

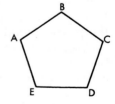

Truncating a Cube

To truncate a solid means to cut off corners. Truncating the regular polyhedrons can produce some of the most impressive and beautiful of the semiregular solids. In general, these have faces of more than one shape but identical corners.

The value of such models in the mathematics class is inherent in their special shapes and properties. But indeed, of even more value is the skill in space perception that comes from imagining certain of these solids formed from truncating others. This activity is suggested with that in mind. Designed for the better student, it can serve as an exciting, challenging exercise in visualization which culminates, rather than begins, with the model.

Begin this exploration by reviewing properties of a cube such as the number of vertices, faces, and edges. Then build on these easy perception problems with more challenging questions such as those that follow. Encourage students to tax their visualization powers to the fullest. Use a model of a cube only as a final verification of the initial answer.

Imagine cutting off each corner of a cube by passing a plane through the midpoints of the three adjacent edges. What results when a cube is truncated this way?

The result is a cuboctahedron. Here is a net that can be used in constructing a model of a cuboctahedron.

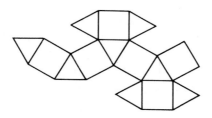

If you cut off the corners of a cube in stages, starting first with very small pieces, this interesting sequence of solids is found. The construction of a complete set of these models would make an interesting special project.

Stage 1:	cube	6 square faces
Stage 2:	truncated cube	6 octagonal and 8 triangular faces
Stage 3:	cuboctahedron	6 square and 8 triangular faces
Stage 4:	truncated octahedron	6 square and 8 hexagonal faces
Stage 5:	octahedron	8 triangular faces

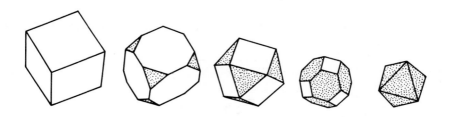

The original 6 faces become octagonal, then square again, and finally disappear. The original 8 corners become triangular, then hexagonal, and finally triangular again. Surprisingly, if you start with the regular octahedron and begin truncating corners, the reverse sequence is formed, and the final solid is again a cube. This describes why the solid in the middle is given the combined name cuboctahedron.

A thorough discussion of this successive truncation of a cube yields a vast amount of visualization experience, some challenging problems to solve, and a glimpse into the subject of semiregular polyhedrons. The mere fact that the equilateral triangle, regular hexagon and regular octagon are part of this process that starts with the squares on a cube illustrates again the beauty of mathematical relationships!

The two visualization activities given thus far were discussed in some detail and each can serve as the source of a classroom lesson on the subject. Similar experiences are noted elsewhere in this chapter. But visualization activities can take the form of short questions to start off the class, to end the week, or as an extra-credit challenge. Here are a few of the many possibilities. Choose from them and use them regularly during the year, not just when studying geometry.

1. Take a long, narrow strip of paper. Carefully tie an overhand knot in it and then crease it flat. What regular polygon is formed?

2. A solid with flat faces has this view for both its top and front. Draw its side view.

3. Show how a solid regular tetrahedron can be cut into two congruent halves. What shape is the intersecting cut?

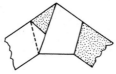

4. Describe the shape of a single solid that will snuggly fit through each of these openings.

This next set deals only with cubes, which can serve as excellent models in a wide variety of visual applications.

5. How many 1 x 1 x 1, 2 x 2 x 2, and 3 x 3 x 3 cubes can be found in this figure?

6. List the six paths possible along the edges of the cube that pass through all eight vertices in going from vertex *A* to vertex *G*.

7. Name the polygon formed when each cube is cut by a plane passing through the midpoints of the three sides shown.

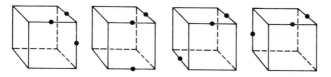

8. Describe the shape of the surface formed by spinning a cube using a major diagonal as an axis.

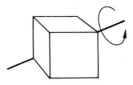

These problems, while easily stated and assigned, can lead to some very creative visual experiences. Remember, encourage students to share and discuss their intuitive ideas, their plans of attack, their diagrams and listings, their models, and their experiments, whether they led them to the correct and complete solution or not.

The two-student worksheets that follow illustrate yet another possible format for visualization experiences. Students should try these first without any models or aids.

Can you spot the cubes?

Many different patterns can be used to form models of cubes. Each pattern must have six squares for faces arranged so that, when assembled, no faces overlap.

1. Study the patterns given here and circle those that you think can be used for a cube. Be careful!
2. Check your answers by cutting out the patterns you are not sure of and trying to assemble them.
3. Copy those that work onto a sheet of graph paper. Then draw as many more as you can find. Remember, count only those that are different patterns, not those that are different positions of the same pattern.

What do you see?

Imagine that this 3 x 3 x 3 cube is painted red and then cut into 27 small 1-inch cubes.

1. Complete the table describing how the 27 small cubes are painted.

NUMBER OF FACES PAINTED	NUMBER OF SMALL CUBES
0	
1	
2	
3	
4	
total	27

2. Suppose that a 4 x 4 x 4 cube was cut into 1-inch cubes the same way. How would the 64 small cubes be painted?

NUMBER OF FACES PAINTED	NUMBER OF SMALL CUBES
0	
1	
2	
3	
4	
total	64

3. If you had no trouble with these, try the same with an n x n x n cube. Be sure that your answers add to n^3.

7.7 MEASUREMENT EXPERIMENTS

Hands-on experiences are a common part of most measurement units taught in the classroom. However, it is still surprising just how many interesting experiments are possible involving measurement and estimation.

Estimation should be an important part of teaching measurement. Try some of these with your class. Bring out the objects and measuring tools only after the students make their estimates.

1. What is the length of a dollar bill in inches?
    (a) 5   (b) 6   (c) 7   (d) 8   (e) 9

2. What is the ratio of the length of a dollar bill to its width?
    (a) 5/3   (b) 6/3   (c) 7/3   (d) 8/3   (e) 9/3

3. How many pennies can be placed on a dollar bill without overlapping the edges of the bill or other pennies?
    (a) 18   (b) 21   (c) 24   (d) 27   (e) 30

The experiments that follow illustrate the wide diversity of measurement applications over various age and ability levels.

EXPERIMENT 1   Angles of a Triangle

Material

A triangular piece of paper

Directions

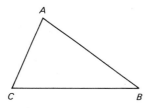 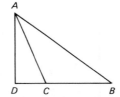 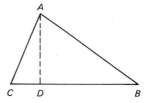

1. Begin with a triangular region *ABC*.

2. Fold vertex *C* onto side *BC* so that the crease passes through vertex *A*.

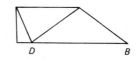

3. The crease *AD* is the altitude through vertex *A*.

4. Fold vertex *A* onto point *D*.
    Fold vertex *C* onto point *D*.
    Fold vertex *B* onto point *D*.

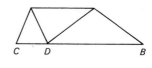

Show that the sum of the measures of the angles *A*, *B*, and *C* is 180.

The three angles of the triangle have been folded onto a straight line. Hence the sum of their measures must be 180.

EXPERIMENT 2   Area of a Triangle

Material

Use the folded triangle from experiment 1.

Directions

1. Note that twice the area of the rectangle is the area of the triangle.

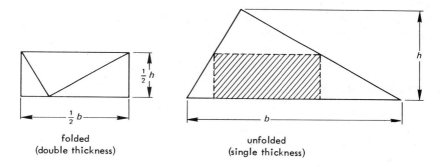

folded
(double thickness)

unfolded
(single thickness)

2. The length of the rectangle is half the base of the triangle. The width of the rectangle is half the height of the triangle. Show that the area of the triangle is $\frac{1}{2}bh$.

Analysis

$$\begin{aligned}
\text{Area of triangle} &= 2(\text{Area of rectangle}) \\
&= 2(l \times w) \\
&= 2\left(\frac{1}{2}b \times \frac{1}{2}h\right) = 2\left(\frac{1}{4}bh\right) = \frac{1}{2}bh
\end{aligned}$$

Many students enjoy field activities in the mathematics class. The *hypsometer* is a simple, useful instrument designed to help determine the height of an object by indirect measurement. It can be made and used by the students themselves with results that can be surprisingly accurate. Adaptable to a variety of levels, it can be effectively used as a supplement to units on measurement, scale drawing, and ratios, or even as an introduction to numerical trigonometry.

The principle of the hypsometer is based on similar figures.

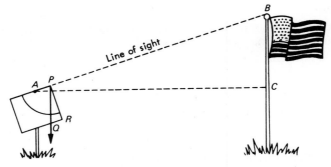

Right triangle *PQR* is similar to right triangle *ABC*. The plumb line determines the vertical line *PQ*; the horizontal line *AC* is determined by sighting with the plumb line aligned to the edge of the hypsometer *PR*. The distance from *C* to the ground (the height of the instrument on level ground) must be added to the height read from the hypsometer to determine the actual height of the flagpole.

EXPERIMENT 3   Hypsometer

Materials

The scale shown, a plumb line, and a pole

Directions

Students can draw the scales themselves or attach a copy supplied by the teacher on a rectangular piece of cardboard. Suspend a plumb line from the upper right-hand corner of the grid as shown and attach it to a pole of convenient height so that it will pivot freely.

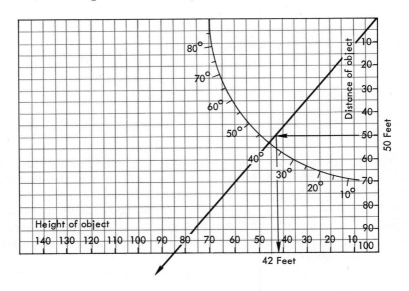

Analysis

With the hypsometer, the height can be read directly from the grid, given the distance to the object. Note how similar triangles are involved in the use of the scales. For the setting shown, the distance to the object is 50 feet. The corresponding height of the object is 42 feet plus the height of the instrument.

EXPERIMENT 4   Volume Relationships

Material

Two regular hexagons of the same size cut from paper

Directions

Fold each hexagon along the three main diagonals and cut along one to the center. Curl up and assemble one as a square pyramid and the other as a regular tetrahedron. Hold them together using paper clips. Now find their volume relationship.

Analysis

All edges of both pyramids have the same length, *e*. So find both volumes in terms of *e* and compare them.

$$V = \frac{1}{3}Bh$$

$$V_T = \frac{1}{3}\left(\frac{\sqrt{3}}{4}e^2\right)\left(\frac{\sqrt{6}}{3}e\right) = \frac{\sqrt{2}}{12}e^3$$

$$V_S = \frac{1}{3}(e^2)\left(\frac{\sqrt{2}}{2}e\right) = \frac{\sqrt{2}}{6}e^3$$

7.8 MORE AIDS AND ACTIVITIES

This section contains activities on the Pythagorean property and spirolaterals. They can add interest and excitement to the classroom scene.

Pythagorean Property

Many different aids can be used in teaching the Pythagorean property. Some serve best for initial investigation and discovery while others are best in review. Some are very formal and exact while others are informal and approximate. Several are illustrated here.

This first model offers convincing support for the property but only for the 3–4–5 right triangle. Similar models can be made for other Pythagorean triples. A modification, drawn on graph paper, allows any positive integers for the legs. The square on the hypotenuse can always be subdivided by this method and its area easily found.

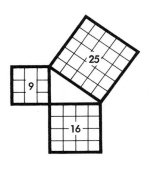

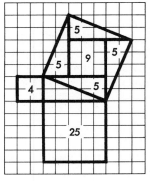

$$3^2 + 4^2 = 9 + 16$$
$$= 25$$

The legs are 3 and 4 and the hypotenuse is 5.

$$2^2 + 5^2 = 4 + 25 = 29$$
$$29 = 9 + 4(5)$$

The legs are 2 and 5 and the hypotenuse is $\sqrt{29}$.

This puzzle offers a challenging hands-on activity for your students. Cut out the squares on the two legs, and subdivide the larger one as shown. Arrange the five pieces so that they just fill the square on the hypotenuse.

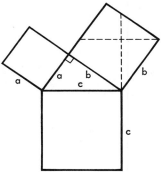

Some time after the puzzle is solved, give it an added twist. Pass out paper squares and ask your students to cut them up and arrange the pieces to form two squares. See how many can transfer what they have learned here to this new situation.

Spirolaterals

Spirolaterals are geometric designs generated from number sequences. They offer interesting opportunities for exploration and discovery and have obvious applications to symmetry and coordinate geometry.

Start with a short number sequence.

Move the first number of units right.
Move the second number down.
Move the third number left.

Move the fourth number up, and so on.

Continue making 90° turns in a clockwise direction, repeating the number sequence as needed.

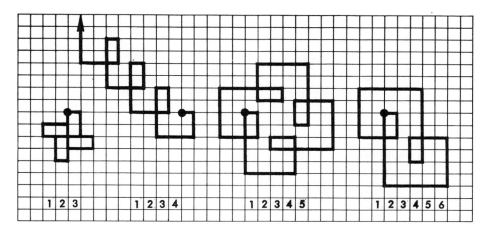

Based on the examples shown, try to guess the properties for the designs generated by these number sequences.

112   1122   11223   112233   1122334   11223344

Which will repeat after two cycles and which after four?
Which will never return to the starting point?

Now try some other number sequences on your own and study the corresponding spirolaterals. A computer program written to generate a spirolateral for each student's name when the letters are typed in can be found in the last section of this chapter. (See page 238.)

7.9 OVERHEAD PROJECTOR IDEAS

Of all the audiovisual equipment available today, none has greater potential in the mathematics classroom than the overhead projector. Its simplicity, versatility, and adaptability can make it an indispensable aid for the teacher. With some imagination and creativity, many striking effects are possible, especially in geometry.

The examples given here take advantage of the overhead projector's ability to show effectively a sequence of separate steps and continuous motion and change.

Construction of a Regular Pentagon

Prepare a transparency in advance showing the separate steps in the construction of a regular pentagon. Students can follow the sequence, but as a teacher, you can refer back to any step at any time.

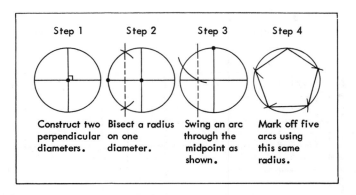

Step 1	Step 2	Step 3	Step 4
Construct two perpendicular diameters.	Bisect a radius on one diameter.	Swing an arc through the midpoint as shown.	Mark off five arcs using this same radius.

Sketching Prisms

The separate steps of a construction sequence can be illustrated impressively using overlays. Start with an initial figure on the base. Put each successive step on a separate transparency. The final result appears when all overlays are in place on the base.

In this example, any step of the sketching process can be referred to at any time.

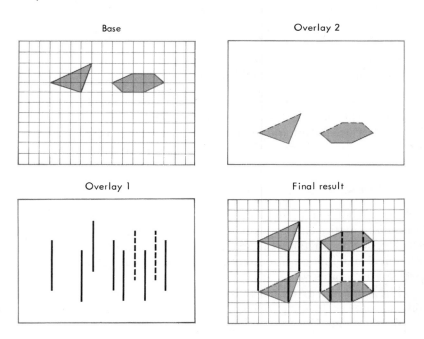

Base

Overlay 2

Overlay 1

Final result

Reading a Protractor

Use a copy machine to reproduce a protractor on a sheet of acetate. Cut two acetate strips with rays drawn on them and attach them to the center point of the protractor with a sewing snap as shown.

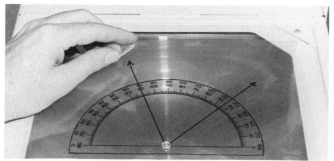

Practice reading the angle size on the correct scale when the initial side is to the right. Then place the initial side to the left and practice reading the other scale. Finally, move both rays off the base line and have students discuss some methods that can be used to find the measure of the angle between them.

Classifying Triangles

Motion is created here using a moveable mask attached to an opaque base with a sewing snap. As the small rectangular mask is turned, the projected triangle passes through various stages of classification. The effect is vivid and striking!

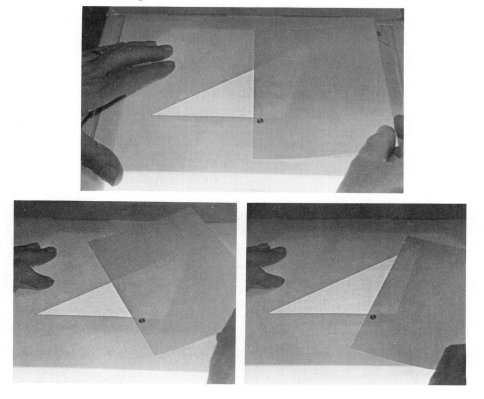

Begin with the moveable mask in the upright position projecting a right triangle. Students will be surprised to see the triangle change shape as the mask is rotated counterclockwise until the triangle disappears entirely. Ask at that point if any other right triangles were formed in the process. Then, as you gradually open up the triangle by rotating the mask clockwise, have the students classify the projected triangle as acute, right, or obtuse. The projected triangle is acute between the two positions of the right triangles. Otherwise, it is obtuse. Now close the figure again. Ask the students to imagine its opening up and count the number of different places where an isosceles triangle is formed. Many students will not see all three positions until the triangles are actually projected.

7.10 CALCULATOR AND COMPUTER APPLICATIONS

Many geometric applications of the calculator and computer can be found in professional journals and commercial sources. The ones noted here are merely illustrative of the diversity possible.

Angles of a Polygon

Most students know the number of degrees in each of the interior angles of an equilateral triangle and a square. Those for a regular pentagon, hexagon, and octagon can easily be found.

equilateral triangle	60° angles
square	90° angles
regular pentagon	108° angles
regular hexagon	120° angles
regular octagon	135° angles

For any regular polygon, first subdivide into triangles. Take the number of triangles, multiply by 180, and then divide by the number of sides in the regular polygon. In general, for a regular n-gon, the number of degrees per angle is

$$\frac{(n - 2)180}{n}$$

But to explore what happens to the size of the angles of a regular polygon as the number of sides increases focuses not on the computation but on the pattern or trend in the results. A calculator facilitates the computation.
Find the number of degrees in the angles of these regular polygons.

10 sided	(144°)
100 sided	(176.4°)
1000 sided	(179.64°)
10,000 sided	(179.964°)

How many sides does a regular polygon have with angles of these sizes?

179°   179.9°   179.99°   179.999°

Try guess-and-test using the calculator, but also look for a shortcut.

Students familiar with LOGO might be the first to see the role the exterior angle plays. Note how regular polygons are generated in this LOGO program using the fact that the exterior angles of a regular *n*-gon contain 360/*n* degrees each.

```
TO POLYGON :N                    ?POLYGON 5
IF :N > 8 [STOP]
HOME
RT 90
REPEAT :N [FD 25 LT 360/:N]
POLYGON :N+1
END
```

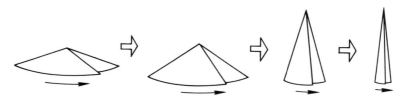

Maximum Volume

The microcomputer is a problem-solving tool to be brought out and used at the right time and place. Teachers need classroom activities that illustrate this point. Some of the most convincing applications come out of problems that start with no thought of this technology in mind. Here is an example.

Give a circular piece of paper of uniform size cut through once to the center to each of the students. Have them curl it up to form a cone of changing radius and height but with a fixed slant height.

The elements of the limit concept should become part of the students' thinking long before they see it in the context of the calculus. Use this model to encourage some informal thinking along these lines.

As the height approaches 0,
  what length does the radius approach?   (slant height)
  what does the volume approach?   (0)
  what does the surface area approach?   (twice area of original circle)

As the length of the radius approaches 0,
  what does the height approach?   (slant height)
  what does the volume approach?   (0)
  what does the surface area approach?   (0)

Now have every student do this estimation activity.

Curl up the cone to the place where it appears to have maximum volume and fasten it with a paper clip at that point.

Have the students hold up their cones so the whole class can see the choices. There are sure to be many, but which has the maximum volume?

There are many approaches to finding the best position for maximum volume. See if any students suggest using a microcomputer.

Assume that the radius of the original paper circle is 8 cm. This program in BASIC computes and prints out volumes in cubic centimeters for all radii from 0 to 8 cm in increments of 0.5 cm.

```
10  PRINT "RADIUS", "HEIGHT", "VOLUME"
20  PRINT
30  FOR R = 0 TO 8 STEP .5
40  LET H = SQR (64 - R * R)
50  LET V = 3.1416 * R * R * H / 3
60  LET H = INT (H * 100 + .5) / 100
70  LET V = INT (V * 100 + .5) / 100
80  PRINT R, H, V
90  END
```

Students can scan the table for the radius that gives the maximum volume and compare it with their own estimates. The answer will be a surprise for most!

RADIUS	HEIGHT	VOLUME
0	8	0
.5	7.98	2.09
1	7.94	8.31
1.5	7.86	18.52
2	7.75	32.45
2.5	7.6	49.74
3	7.42	69.9
3.5	7.19	92.28
4	6.93	116.08
4.5	6.61	140.26
5	6.24	163.49
5.5	5.81	184.03
6	5.29	199.49
6.5	4.66	206.34
7	3.87	198.73
7.5	2.78	163.98
8	0	0

How can the program now be modified for a more precise answer by incrementing in units of 0.1 cm for radii from 6 to 7 cm?

Spirals

Do not overlook the graphic capabilities of the microcomputer when teaching geometry. It has extraordinary potential in this area. These next three programs written in LOGO illustrate the point on the topic of spirals.

SPIRAL1 generates a rectangular spiral from a pair of segments at right angles to each other. The third line of the procedure draws an initial vertical and horizontal line segment using the first two input values. Recursion is initiated in the fourth line. This causes the process to cycle over and over increasing the lengths of the segments each time using the third input. The recursion stops when one of the limits imposed in the first two lines is reached.

Students can display results using various input values. They can also study exactly what is happening to these values each time the recursion is called. Finally, they can modify the recursion line and explore the effects on the spiral. A run for SPIRAL1 is shown here for inputs of 5 10 5.

```
TO SPIRAL1 :A :B :C
IF :A >100 [STOP]
IF :B >200 [STOP]
FD :A RT 90 FD :B RT 90
SPIRAL1 :A+:C :B+:B/:A*:C :C
END
```

?SPIRAL1 5 10 5

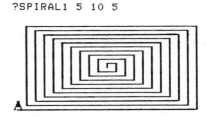

SPIRAL2 generates a spiral from successive right triangles where the hypotenuse of one right triangle becomes a leg of the next. Applications of this spiral in the study of the Pythagorean theorem and square roots are obvious.

?SPIRAL2

```
TO SPIRAL2
MAKE "S 0
REPEAT 17[FD 20 * SQRT (:S) BK 20 *SQRT
(:S) RT 90 FD 20 SETHEADING TOWARDS [0
0] MAKE "S :S +1]
END
```

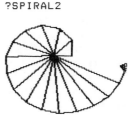

SPIRAL3 draws a spiral based on the Golden Ratio (see page 185). The initial rectangle and all subsequent rectangles are golden rectangles. Successive arcs are drawn in squares formed from these rectangles.

```
TO SPIRAL3 :S
IF :S <1 [STOP]
REPEAT 2[FD :S RT 90 FD :S * 1.61803 RT
90 ]
REPEAT 90[FD 2 * :S * 3.14 /360 RT 1 ]
SPIRAL3 :S*0.61803
HT
END
```

?SPIRAL3 80

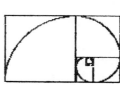

Spirolaterals, first discussed on page 230, are a special form of spirals generated from number sequences. If the letters A through Z are coded with digits as shown, then spirolaterals can be drawn for names.

A	B	C	D	E	F	G	H	I	J	K	L	M	N	O	P	Q	R	S	T	U	V	W	X	Y	Z
1	2	3	4	5	6	7	8	9	1	2	3	4	5	6	7	8	9	1	2	3	4	5	6	7	8

Using this code, students can draw their own individual spirolaterals on graph paper. An exhibit of the spirolaterals for all the names in the class would make an impressive bulletin board display.

This program, written in BASIC for the high resolution graphics of the APPLE microcomputer, draws the corresponding spirolateral for any name that is entered. A scaling factor of 3 is used in coding the letters to generate a figure of appropriate size for the screen.

```
5   REM     *SPIROLATERALS*
10  REM  *INPUT NAME AND GET LENGTH*
20  INPUT "WHAT IS YOUR NAME? ";NA$
30  LE =  LEN (NA$)
35  DIM L$(LE)
40  FOR K = 1 TO LE
50  L$(K) =  MID$ (NA$,K,1)
60  NEXT K
65  X = 140:Y = 80
66  K = 1
70  HGR
80  HCOLOR= 3
90  LET S = 1
100   GOSUB 2000
110   LET F = (S / 4 -  INT (S / 4)) * 4
120   IF F = 1 THEN P = W:Q = 0: REM  *MOVE RIGHT*
130   IF F = 2 THEN P = 0:Q = W: REM  *MOVE DOWN*
140   IF F = 3 THEN P =  - W:Q = 0: REM  *MOVE LEFT*
150   IF F = 0 THEN P = 0:Q =  - W: REM  *MOVE UP*
160   IF X + P > 279 OR X + P < 0 OR Y + Q > 159 OR Y + Q < 0
          THEN  END : REM  *OFF SCREEN*
170   HPLOT X,Y TO X + P,Y + Q
180   X = X + P:Y = Y + Q
190   IF K / LE =  INT (K / LE) AND X = 140 AND Y = 80
          THEN  END : REM  *BACK HOME*
200 K = K + 1: IF K = LE + 1 THEN K = 1
205   LET S = S + 1
210   GOTO 100
2000    IF L$(K) = "A" OR L$(K) = "J" OR L$(K) = "S" THEN W = 3
2010    IF L$(K) = "B" OR L$(K) = "K" OR L$(K) = "T" THEN W = 6
2020    IF L$(K) = "C" OR L$(K) = "L" OR L$(K) = "U" THEN W = 9
2030    IF L$(K) = "D" OR L$(K) = "M" OR L$(K) = "V" THEN W = 12
2040    IF L$(K) = "E" OR L$(K) = "N" OR L$(K) = "W" THEN W = 15
2050    IF L$(K) = "F" OR L$(K) = "O" OR L$(K) = "X" THEN W = 18
2060    IF L$(K) = "G" OR L$(K) = "P" OR L$(K) = "Y" THEN W = 21
2070    IF L$(K) = "H" OR L$(K) = "Q" OR L$(K) = "Z" THEN W = 24
2080    IF L$(K) = "I" OR L$(K) = "R" THEN W = 27
2090    RETURN
```

Here are some of the types of runs you can expect.

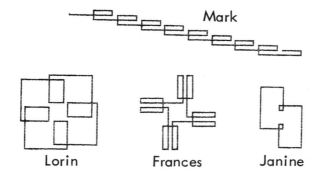

Mark

Lorin   Frances   Janine

EXERCISES

1. An unlimited number of squares and rectangles that measure 2 x 2, 2 x 4, 2 x 6, 4 x 4, 4 x 6, and 6 x 6 are available. How many different-sized rectangular prisms can be constructed from them using one rectangle per face?

2. Refer to the figure of the tangram puzzle given on page 196. If square ABCD measures 2 inches on each edge, find the area and perimeter of trapezoid MBOP. Name two other trapezoids congruent to it.

3. An 8 1/2 x 11-inch sheet of paper is curled up end-to-end in two different ways to form two different cylinders. Find the volume of each.

4. Twelve congruent square pieces of paper are cut out, six red and six green. How many different-colored models of a cube can be made using any six of the squares?

5. A cone is formed from a circular piece of paper cut through to the center. When curled up in different ways, different cones are formed all with the same slant height. Draw graphs to show the relationship between the radius and the height, the radius and the volume, and the radius and the total surface area.

6. Archimedes proved that the volume relationship among a cone, sphere, and cylinder of the same radius and height is 1 to 2 to 3. Prove this relationship algebraically.

7. Show how Euler's formula $V + F = E + 2$ applies to a prism and a pyramid with an n-gon as bases.

8. Find the number of vertices, faces, and edges in each of the five regular polyhedrons. Look for shortcuts in doing the counting.

9. Five 1-cm cubes are joined edge-to-edge in all possible ways. How many different figures are possible?

10. Prove that only five regular polyhedrons are possible.

11. A 32-inch string loop is used to draw an ellipse around two thumbtacks 12 inches apart. Write the equation of the ellipse formed if the origin is at the midpoint of the segment connecting the thumbtacks and the axes are marked in inches.

12. A triangle is cut from paper and its vertices folded to the midpoints of the opposite sides. What condition must be imposed on the original triangle such that the folded model can be used to form a tetrahedron?
13. Count the number of vertices, faces, and edges of a cuboctahedron.
14. Find the volume and surface area of the cuboctahedron formed from a 4-inch cube by cutting off its corners with planes passing through the midpoints of the adjacent sides.

ACTIVITIES

1. Use a string knotted at 1-foot intervals to lay out a 10-foot square as the ancient Egyptians might have done.
2. Fold a square piece of paper in half vertically. Explore what happens when you try to cut the result vertically such that the ratio of the perimeters of the resulting rectangles, smaller to larger, is 1 to 2.
3. Construct a model of a cube by the plaiting technique described on page 210.
4. Cut some 3 x 3 squares from graph paper. Divide them into two congruent halves using polygonal paths joining intersection points on the grid. Thirteen different-shaped congruent halves are possible. Try to find them all.
5. Find the 10 concave polygons that can be formed from the paper-folding activity described on page 212 by folding over or under on any crease.
6. Describe in detail an experiment involving a ruler, string, and some circular objects that students might follow in discovering an approximate value for π.
7. Identify several locus problems that can be easily illustrated with simple aids in the geometry classroom.
8. Show how to paper fold a regular hexagon and a regular octagon.
9. See how many different circle relationships can be demonstrated using paper-folding illustrations.
10. Fold a parabola, ellipse, and hyperbola on waxed paper that can be shown on the overhead projector following the methods shown on page 218.
11. Develop a paper-folding experiment that can be used to illustrate how to find the area of a parallelogram. Use the same method to illustrate that the midpoint of the hypotenuse of a right triangle is equidistant from the three vertices.
12. Prepare a set of activity cards that a student can use with a geoboard to develop a specific geometric formula or property.
13. Develop an experiment that students can follow in discovering the pattern relating the number of diagonals to the number of vertices in a polygon.
14. Plan a field activity for a junior high class that would make use of a hypsometer.
15. Make a set of models showing the various stages in the truncating of a cuboctahedron as shown on page 221. Start with a 4-inch cube.
16. Describe five different junior high classroom activities that center around the student's construction or use of a model of a cube.
17. Prepare a lesson plan on Euler's formula for an eighth-grade class. Pay specific attention to the motivation of the lesson, the use of classroom models, individual student involvement, and an appropriate assignment.

18. Construct a transparency that can be used to review the formula for the area of a trapezoid, $A = \frac{1}{2}h(a + b)$. Show the steps to follow in a flow chart and use masked disks to select various values of h, a, and b.

19. Discuss how a rectangular grid projected on a chalkboard can be used to teach congruency.

20. Develop a transparency sequence that can be used when teaching the Pythagorean theorem.

21. Write for a catalog from one of the suppliers of materials and models for the mathematics classroom. Then use it to identify and describe some commercially available aids for geometry, indicating their potential use.

22. Find a source and price for a commercially made wooden cone that can be dissected and used to illustrate the conic sections.

23. Cut a paper square into acute triangles. This problem is more challenging than it may appear at first. The key to the solution rests on the extension of the theorem in geometry concerning angles inscribed in a semicircle.

READINGS AND REFERENCES

1. The 1987 Yearbook of the National Council of Teachers of Mathematics is *Teaching and Learning Geometry, K-14.* Prepare a written summary of two of its chapters.

2. Read *Flatland* by Edwin Abbott, New York: Dover Publications, Inc., 1952. This is a reprint of an original text dating back to the late 1800's. Prepare a plan for a short presentation to a junior high school mathematics class on his description of life in a two-dimensional world.

3. Read Chapter 4 titled "Assorted Geometries—Plane and Fancy" in *Mathematics and the Imagination* by Edward Kasner and James Newman, New York: Simon and Schuster, 1967. List some of the key ideas given on the fourth dimension and on non-Euclidean geometry that could be presented to a high school geometry class.

4. Read Chapters 10 and 11 in *Mathematics in Western Culture* by Morris Kline, New York: Oxford University Press, 1953. Prepare a report on the impact of Renaissance painting and perspective on the creation of projective geometry in the seventeenth century.

5. Read Chapter 3 in *What Is Mathematics?* by Richard Courant and Herbert Robbins, New York: Oxford University Press, 1941. Prepare a report on the relationship between geometric constructions and algebra.

6. Read Chapter 16 in *Men of Mathematics* by E. T. Bell, New York: Simon and Schuster, 1965. Prepare a report on the mathematician Bell calls "the Copernicus of Geometry."

7. One of the classics on the construction of mathematical models is *Mathematical Models* by H. M. Cundy and A. P. Rollett, New York: Oxford University Press, 1951. Read Chapter 3 and construct several of the Kepler-Poinsot, Archimedean, and stellated Archimedean polyhedra described therein.

8. Read Chapter 4 in *Mathematical Snapshots* by H. Steinhaus, New York: Oxford University Press, 1969. Prepare a display of at least five of the homogeneous tessellations described in the chapter.

9. Prepare a report on some of the history surrounding the regular polyhedrons and famous mathematicians, such as Pythagoras, Plato, Euclid, Archimedes, and Kepler, who studied them.

10. Read and report on President Garfield's proof of the Pythagorean theorem.

11. Geometry textbooks are likely to have major similarities as well as striking differences. Compare two current geometry textbooks recording and commenting on what you see as their basic differences.

12. Read the September 1985 special geometry issue of the *Mathematics Teacher* of the National Council of Teachers of Mathematics. Identify and comment on the article that you find most interesting.

13. The National Council of Teachers of Mathematics offers a variety of publications on geometry. Prepare a report on one of those listed here.

> *Four-Dimensional Geometry—An Introduction* by Hess, 1977
> *How to Enrich Geometry Using String Designs* by Pohl, 1986
> *Mathematics Through Paper Folding* by Olson, 1975
> *Polyhedron Models for the Classroom* by Wenninger, 1975

14. The *Mathematics Teacher* of the National Council of Teachers of Mathematics contains numerous articles on microcomputers and programs as they relate to the teaching of geometry. Identify three articles of your choice citing title, author, and date. Read them through, run the programs included, and evaluate their potential effectiveness.

15. Read Chapter 3 titled "Geometry Concepts" in *Using Computers in Mathematics* by Gerald Elgarten, Alfred Posamentier, and Stephen Moresh, Menlo Park, California: Addison-Wesley Publishing Company, 1983.

Classroom Aids and Activities: Probability and Statistics

Chapter 8

Probability and statistics are playing an ever-increasing role in society. This has come about, in part, because of the vast increase in data-handling and graphics capabilities of the computer. At the same time, educators are asking that the priorities in the instructional program be changed to include increased emphasis on such activities as collecting, organizing, presenting, and interpreting data, as well as on using collected data to draw inferences and make predictions. This chapter offers some suggestions on classroom aids and activities related to these recommendations.

8.1 MOTIVATING PROBABILITY AND STATISTICS

The subjects of probability and statistics tend by their very nature to be more motivating, concrete, and manipulative oriented than most other topics in mathematics. Related activities and experiments on these topics are more commonplace in the curriculum than most others, but the pressures of time often lead the teacher to omit the units on probability and statistics entirely.

Use Actual Real-World Data

When computing and comparing the mean, median, and mode, look for examples that offer more than just the facts themselves. Consider the following table of the monthly rainfall in Bombay:

MONTHLY RAINFALL IN BOMBAY

MONTH	J	F	M	A	M	J	J	A	S	O	N	D
RAINFALL IN INCHES	0	0	0	0	1	21	27	16	12	2	0	0

Would you say the weather in Bombay is dry or wet?

Does the mean of 6.6 inches, the median of 0.5 inches, or the mode of 0 inches best describe the monthly rainfall? Or is some other description needed here?

How do these data compare with those of your own town or city?

Encourage Estimation

Have each student guess the diameter of a penny to the nearest eighth of an inch by choosing from these estimates:

$$3/8 \quad 1/2 \quad 5/8 \quad 3/4 \quad 7/8$$

Let them use the results to make a probability statement about such guesses. Then have them measure a penny to see which estimate is best.

Allow Students to Create Problems

Bring in some objects and allow students to create their own probability problems from the situations posed. Let others in class offer answers. Here are some examples of items that can be used:

7 pennies and 5 nickels in 20 coins
3 E's and 5 other vowels on 10 lettered cards
4 jacks, 4 queens, and 4 kings in a deck of 52 cards

Choose Activities That Involve All Students

Ask each student to bring in three pictures of the same size, each cut in half vertically. Turn the pieces over, mix them up, and have another student select two pieces at random. What is the probability that the two pieces match?

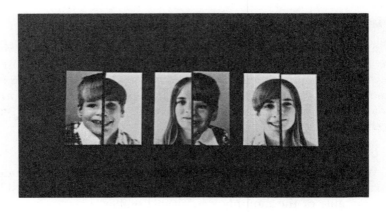

First, have every student conduct the experiment in class, record the results, and use them to compute an experimental probability. Next, let each student see how many different pairs of pieces are possible (15) and how many match (3). Finally, use these numbers to compute the theoretical probability (1/5).

For some classes, you may want to facilitate their listing of choices by suggesting matched halves be labeled A and a, B and b, and C and c.

AB   AC   *Aa*   Ab   Ac   BC   Ba   *Bb*   Bc   Ca   Cb   *Cc*   ab   ac   bc

Make Use of Familiar Aids and Models

Dominoes are not only useful teaching aids in the early grades for showing number relationships and fraction concepts, but they can make effective teaching aids in elementary probability as well.

> A domino is chosen at random from a complete set. What is the probability that it contains a 5?

For elementary students, bring in a complete set of dominoes, lay them out, and let the students count the total number (28) and those with fives (7).

In a more typical situation, show a single domino or two to establish the criteria for making and distinguishing dominoes. Cardboard models, cut out and punched, can be very effective on the overhead projector. Then challenge each student to make a complete list of all possibilities. Encourage a systematic listing, such as this:

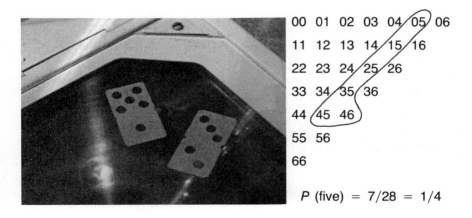

00 01 02 03 04 05 06

11 12 13 14 15 16

22 23 24 25 26

33 34 35 36

44 45 46

55 56

66

P (five) $= 7/28 = 1/4$

At another level, extend the problem to sums and differences on dominoes. Construct frequency distributions. Compare and contrast the corresponding histograms.

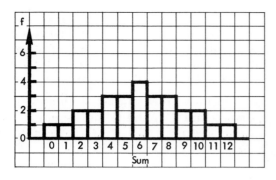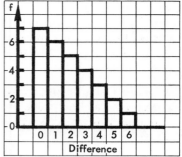

What sum and difference would you expect on a randomly selected domino?

Select random samples and find the average sum and average difference. Compare these empirical values with the actual expected values.

8.2 COUNTING AND PROBABILITY

In its simplest form, theoretical probability begins with a discussion of counting techniques. Counting is at the same time both one of the most trivial and most complex mathematical skills. The use of concrete models and manipulatives with some of the simpler counting problems can give the student a practical, geometric, visual approach that can later on be applied to more involved counting situations and probability problems.

Lettered Cards

Letter some 3 x 5 file cards to use as an aid in discussing counting techniques. Use the name of your school, a student's name, or your own, and build some counting and probability problems around it. The word "STATISTICS" is used here as an illustrative example.

Place each letter on a separate card. Use the cards to explain how each probability is found from a single random drawing.

S T A T I S T I C S

$P(A) = 1/10$ $P(S \text{ or } T) = 6/10$
$P(S) = 3/10$ $P(\text{consonant}) = 7/10$

Use the cards to contrast sampling with and without replacement. Two letters are drawn. What is the probability that both are S's?

$$P(2\ \text{S's}) = \frac{3}{10} \times \frac{3}{10} = 9/100$$

$$P(2\ \text{S's}) = \frac{3}{10} \times \frac{2}{9} = 6/90 = 1/15$$

With replacement of the first S before the second is drawn

Without replacement of the first S before the second is drawn

Use cards to contrast permutations and combinations. How many permutations and combinations of 2 of the 5 letters in the word "COUNT" are possible?

$$_5P_2 = 5 \times 4 = 20$$

$$_5C_2 = \binom{5}{2} = \frac{5 \times 4}{2 \times 1} = 10$$

There are 20 permutations or *orderings* possible, but there are only 10 combinations or *choices* possible. Each pair of letters constitutes one combination or choice but two permutations or orderings.

Use cards to count more complicated letter arrangements. In how many ways can the eight letters in "EIGHTEEN" be arranged?

EIGHTEEN

All possible arrangement

$$\frac{8!}{3!} = 8 \times 7 \times 6 \times 5 \times 4 = 6720$$

For all E's together, replace the three E cards with a single one and count the orderings possible with those six cards. Then simply replace the one E card with the three E's for the 8-letter arrangements with the three E's together.

EIGHTN

Arrangements with E's together

$$6! = 6 \times 5 \times 4 \times 3 \times 2 \times 1 = 720$$

What is the probability that a random arrangement of the letters in "EIGHTEEN" have all E's together?

Numbered Cubes

Numbered cubes or dice are especially useful teaching aids. Using blank cubes that can be made or purchased commercially and adhesive labels, number cubes can be constructed with any number configuration.

Find an empirical estimate to each probability first and then do the analysis with a sample space for the theoretical probability. Here is an example.

What is the probability of rolling a sum of 7 with a pair of number cubes each marked 1, 2, 3, 4, 5, 6 on its faces?

Roll a pair of dice 100 times, recording the sum for each roll. Divide the number of times a sum of 7 occurs by 100 for the experimental probability.

Then use a sample space to analyze all possible equally likely results. Sums are shown inside the 6 × 6 grid shown on the following page.

SECOND DIE

FIRST DIE

	1	2	3	4	5	6
1	2	3	4	5	6	**7**
2	3	4	5	6	**7**	8
3	4	5	6	**7**	8	9
4	5	6	**7**	8	9	10
5	6	**7**	8	9	10	11
6	**7**	8	9	10	11	12

$P(7) = 6/36 = 1/6$

For variety, use a set of dice consisting of a regular tetrahedron, hexahedron, octahedron, dodecahedron, and icosahedron numbered from 1 to 4, 1 to 6, 1 to 8, 1 to 12, and 1 to 20 respectively. On a given die, each face is as likely to occur as any other, but the probability of a given result changes from die to die.

$$
\begin{aligned}
P(3 \text{ on tetrahedron}) &= 1/4 \\
P(3 \text{ on hexahedron}) &= 1/6 \\
P(3 \text{ on octahedron}) &= 1/8 \\
P(3 \text{ on dodecahedron}) &= 1/12 \\
P(3 \text{ on icosahedron}) &= 1/20
\end{aligned}
$$

What is the probability of rolling a sum of 7 with the tetrahedron and octahedron?

OCTAHEDRON

TETRAHEDRON

	1	2	3	4	5	6	7	8
1	2	3	4	5	6	**7**	8	9
2	3	4	5	6	**7**	8	9	10
3	4	5	6	**7**	8	9	10	11
4	5	6	**7**	8	9	10	11	12

$P(7) = 4/32 = 1/8$

There are 46,080 different ways of rolling all five of the polyhedron dice at one time.

$$4 \times 6 \times 8 \times 12 \times 20 = 46{,}080$$

The chance of rolling a sum of 7 with these five dice is only 15 in 46,080 for a probability of 1/3072. A computer simulation for rolling these five dice can be found on page 271.

Coins

What is the probability that when two coins are tossed, both fall heads? Repeated tosses emphasize the notion of probability *in the long run*. As more trials are performed the computed experimental probabilities tend to level off at the theoretical value. This is vividly illustrated by the line graph and by the cumulative fractions and percents in the table.

Students first guess at the answers. Then, tossing two coins at a time, they compute the ratio of successes (2 heads) to trials and the corresponding cumulative percents. These percents are then plotted on a graph for each of 20 successive trials. Typical results might look like this:

Toss	1	2	3	4	5	6	7	8	9	10	11	12	13	14	15	16	17	18	19	20
Two Heads	X						X						X		X					X
Not Two Heads		X	X	X	X	X		X	X	X	X	X		X		X	X	X	X	
Rate of Successes To Total	$\frac{1}{1}$	$\frac{1}{2}$	$\frac{1}{3}$	$\frac{1}{4}$	$\frac{1}{5}$	$\frac{1}{6}$	$\frac{2}{7}$	$\frac{2}{8}$	$\frac{2}{9}$	$\frac{2}{10}$	$\frac{2}{11}$	$\frac{2}{12}$	$\frac{3}{13}$	$\frac{3}{14}$	$\frac{4}{15}$	$\frac{4}{16}$	$\frac{4}{17}$	$\frac{4}{18}$	$\frac{4}{19}$	$\frac{5}{20}$
Per Cent of Successes	1.000	.500	.333	.250	.200	.166	.285	.250	.222	.200	.181	.166	.230	.214	.266	.250	.235	.222	.210	.250

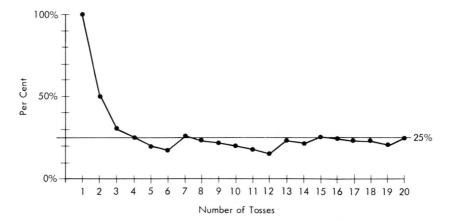

The original guesses can then be compared with the collected data and modified, based upon the experimental evidence.

Students may first guess the probability to be $\frac{1}{3}$, arguing that the coins can fall 2 heads, 1 head, or 0 heads. However, the correct answer is $\frac{1}{4}$, since only one of these four equally likely possible outcomes is a success.

HH   HT   TH   TT

The greater the number of repetitions, the more likely the graph will tend toward this value.

Of course, the assumptions are made throughout the experiment that these are unbiased coins randomly tossed. If these assumptions are in doubt, one might well find the experimental probability the better predictor.

Checkerboards

Pascal's triangle can be developed through the idea of a random walk by a checker on a board. Start with the checker as shown below, and repeatedly toss a coin. If the coin falls heads, move down to the left and if it falls tails, move down to the right. The path of the checker can be thought of as a random walk. The checker moves one square at a time diagonally down to either the left or the right. By counting the number of different ways it can move to various positions on successive rows, the numbers in Pascal's triangle are generated.

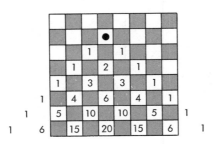

The sixth row of Pascal's triangle has a sum of 2^6 or 64. However, some moves extend beyond the board. Suppose the checker must stay on the board and the coin is tossed only when a choice of moves exist. Then the probabilities of reaching the four white squares on the bottom row, starting from the position shown, would be 14/54, 20/54, 15/54, and 5/54 respectively, left to right.

The numbers in Pascal's triangle are combinations and can readily be applied to the binomial expansion. The numbers in the third row represent the coefficients in the expansion of $(H + T)^3$.

$$(H + T)^3 = 1H^3 + 3H^2T + 3HT^2 + 1T^3$$

Each term also identifies the different random walks to that position.

HHH	HHT	HTT	TTT
HTH	THT		
THH	TTH		

8.3 PROBABILITY EXPERIMENTS

Experiments in probability can be exciting and interesting to students with a wide range of ages and abilities. The element of chance and doubt, the close association with familiar games, and the active involvement of individual students all help to provide motivation. Here is an example of an experiment that might be used in class to introduce the topic.

EXPERIMENT 1   Random Walks

Material

Rulers, paper, and coins

Directions

Each student draws a number line from -5 to 5 and then, starting at 0, moves one unit at a time to the right or left depending upon whether a tossed coin lands heads or tails. The first to reach -5 or 5 tossing his or her own coin wins.

Analysis

Reaching a 5 in five tosses requires 5 heads while reaching a -5 requires 5 tails. The chances of all heads are 1 in 32, as they are for all tails. So if you allow just 5 tosses, the chances are 1 in 16 that a student will reach -5 or 5. In an average-size class, then, at least one student should be expected to win in just 5 tosses.

In 5 tosses, $P(5) = 1/32$, $P(-5) = 1/32$, and $P(5$ or $-5) = 1/16$.

An interesting follow-up is to find how many in class landed at each number after 5 tosses and compare these results with those from a list of all 32 ways in which the 5 tosses could fall.

PROBABILITIES

1 way to stop at -5

TTTTT

$\dfrac{1}{32} = .03125$

5 ways to stop at -3

TTTTH   TTटHT   TTHTT   THTTT   HTTTT

$\dfrac{5}{32} = .15625$

10 ways to stop at -1

TTTHH   TTHTH   THTTH   HTTTH   TTHHT
THTHT   HTTHT   THHTT   HTHTT   HHTTT

$\dfrac{10}{32} = .31250$

10 ways to stop at   1

HHHTT   HHTHT   HTHHT   THHHT   HHTTH
HTHTH   THHTH   HTTHH   THTHH   TTHHH

$\dfrac{10}{32} = .31250$

5 ways to stop at   3

HHHHT   HHHTH   HHTHH   HTHHH   THHHH

$\dfrac{5}{32} = .15625$

1 way to stop at   5

HHHHH

$\dfrac{1}{32} = .03125$

$\overline{\hphantom{xxxxxxx}}$
1.00000

Many classroom experiments can be designed to estimate probabilities that also can be computed mathematically.

Coin tossing:	Heads on a single toss	1/2
	Two heads on two tosses	1/4
	Three alike on three tosses	1/4
Cutting a deck of cards:	Red	1/2
	Diamonds	1/4
	Face card (J, Q, K)	3/13
Rolling dice:	Even number on a single die	1/2
	Two alike with a pair of dice	1/6
	7 or 11 with a pair of dice	2/9

Some probability questions require predictions based on experimental data and the analysis of the results. These yield empirical rather than theoretical probabilities.

What is the probability that a tack lands point up when dropped? In this experiment the probabilities of various tacks falling point up are estimated by experimentation.

EXPERIMENT 2   **Tossing Thumb Tacks**

Material

Thumbtacks of 3 different kinds

Directions

Three different types of thumbtacks are selected. Students first study their construction and try to guess which is most likely to fall point up and which least likely.

Ten tacks of one type are dropped on a flat surface 10 different times. Then the probability that a single tack would land point up is estimated using this ratio:

$$P(\text{up}) = \frac{\text{number falling point up}}{100}$$

The experiment is repeated with the other two tacks and then final results are compared with the original guesses.

Analysis

Experimentation is the best way to estimate probabilities here since no equivalent simple mathematical model can be designed. Some of the characteristics that increase the probability of landing point up are larger heads, flatter heads, heavier heads, and shorter points.

Many problems require much more than just counting and experimenting. Problem-solving skills come into play along with other aspects of the mathematics curriculum. This next experiment ties visualization and geometry to probability.

EXPERIMENT 3   Tossing Pennies on Squares

Material

Pennies and a large sheet of paper ruled in a $1\frac{1}{2}$-inch grid

Directions

Students first guess at what the probability will be that a penny, randomly tossed on the grid, will not fall on a line. Next, 10 pennies are tossed on the grid 10 separate times. The number n of pennies that do not fall on a line is counted and compared with the total number of tosses.

$$P = \frac{n}{100}$$

Answers are then compared with original guesses.

Analysis

A theoretical probability can be determined by comparing areas. Assume that the center of the penny lands anywhere within a given square. The radius of a penny is $\frac{3}{8}$ inch, so if its center lies more than $\frac{3}{8}$ inch from any side of the square, the penny will not be on a line. Hence the only possible location for the center of the penny such that the penny itself is not on a line would be within a small $\frac{3}{4}$-inch square. Compare this area to the total for the original square for the correct probability.

$$P(\text{not on line}) = \frac{\frac{3}{4} \times \frac{3}{4}}{1\frac{1}{2} \times 1\frac{1}{2}} = \frac{1}{4}$$

Center inside small square

Penny inside large square

What is the probability that a penny does fall on a line of the ruled grid?

Here is a famous problem which was first presented by Count Buffon in the 1700s. Needles are dropped in a special way on a ruled surface such that the computed probability that one falls on a line gives an estimate involving π.

EXPERIMENT 4 Buffon's Needle Problem

Material

Toothpicks cut to a 1-inch length
A surface ruled with lines 2 inches apart

Directions

The toothpicks are randomly and repeatedly dropped from a reasonable height onto the ruled surface. The number falling on a line are counted along with the total number dropped. Their ratio should approximate $1/\pi$.

$$\text{To 10 places} \quad \pi = 3.1415926535 \ldots$$
$$1/\pi = .3183098861 \ldots$$

The latter value is always approximated as long as the distance between the lines is twice that of the length of the toothpicks dropped. Increasing the number of trials tends to improve the approximation.

Analysis

A detailed analysis of this problem requires calculus but yields the *exact* result, $1/\pi$.

What are the chances that at least two people in a crowd have the same birthday?

Surveys give data for initial experimental estimates to a problem that has an exact mathematical analysis.

EXPERIMENT 5 The Birthday Problem

Material

People to poll for birth dates and reference sources

Directions

The problem is first carefully discussed. Students should guess what they think are the chances of a duplication with groups of say 15, 25, and 60. Next various groups are surveyed and records kept of the number of duplicate birthdays found in each group. The class itself should be surveyed first. Then fellow students, teachers in the school, parents, or families in the neighborhood can be surveyed along with other recorded data such as the birthdays of all the presidents of the United States. Finally, modifications on the original estimates can be made based upon the collected data.

Analysis

Omitting leap year's February 29, there are 365 possible birth dates. With just two people, the chances are 1 in 365 that they have the same birthday. However, with 3 people, the chances are greater since *any* two could have the same date. The greater the crowd, the greater the chance of a duplication of birth dates.

Here are the surprising results for various group sizes.

SIZE OF GROUP	15	18	23	40	60
Chances of a duplication	better than 1 in 4	better than 1 in 3	better than 1 in 2	better than 9 in 10	almost certain
Approximate probability	1/4	1/3	1/2	9/10	near 1

For *n* people, the exact probability of at least one duplication can be found using the formula

$$1 - \frac{365}{365} \cdot \frac{364}{365} \cdot \frac{363}{365} \cdots \frac{365 - n + 1}{365}$$

Computing these probabilities can be an interesting exercise.
See page 273 for a further discussion of this and similar problems using a microcomputer.

8.4 STATISTICAL ACTIVITIES

Statistical activities in the classroom involve collecting, organizing, presenting, and interpreting data. They also involve drawing inferences from the data collected. Thus the important skills of listing, summarizing, graphing, and predicting are exercised and enhanced along with computational skills.

Estimating Length

Cut off a piece of string, hold it up for only a few seconds for the class to see, and then have the students write down their estimates of its length to the nearest inch. The activity calls for rough, quick estimates, so show the string for only a very short time before putting it away.

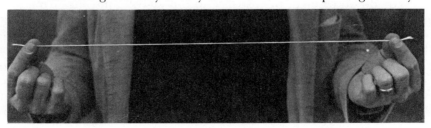

Have the class collect and tabulate the data in a frequency distribution. Have students draw a histogram of the results and comment on its shape. Let them compute the mean, median, and mode and discuss their merits as a typical estimate. Using these collected results, have the class make a final prediction of the actual length of the string. Encourage discussion by asking how confident they are in the prediction. Finally, produce the string and let the class measure it and interpret the comparison or contrast between their individual estimates, the class prediction, and the actual length.

Here are the combined results of several actual class experiments with the piece of string shown in the photo above. What would you guess was the actual length of the string based on these estimates?

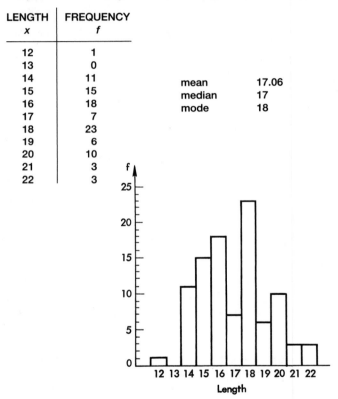

LENGTH x	FREQUENCY f
12	1
13	0
14	11
15	15
16	18
17	7
18	23
19	6
20	10
21	3
22	3

mean   17.06
median 17
mode   18

To add an element of suspense to this activity, put the string in an envelope after its first showing, seal the envelope, and give it to a student for safekeeping. Don't open it up to measure the string for a day or two. Then, when the envelope is opened, make a big fuss about it. The extra fanfare will be well worth it. Not only will it provoke attention and excitement, but it will help to build up to the surprise you can almost certainly expect. For the data shown, the string length was 22 inches! Most students tend to significantly underestimate lengths.

Comparing Heights and Arm Spans

Have students in pairs measure to the nearest inch their heights and their arm spans, fingertip to fingertip with arms outstretched. Let them collect the data and tabulate each set in a frequency distribution before computing the two means. Then compare the two means and see what inferences they can draw.

The average height should be essentially the same as the average arm span.

See page 269 for sample data on this problem.

Predicting Letter Frequencies

What letters are most frequently used in written material? Ask this question of your class and see how students recommend finding the answer. Some will suggest searching through references on the topic while others will suggest sampling written material. An ongoing study of this question can lead to some very interesting statistical experiences.

Begin by having each student guess the three letters used most often. Tabulate the results for the class in a frequency distribution. Discuss the degree of consistency that appears, finding the percent for the top five letters.

Next, have each student sample 500 successive letters from some randomly chosen portion of written material, recording the frequency and percent for each letter. Finally, see how the results compare, first with his or her own initial guesses, second with the pooled results of the class, and third with these percents for ordinary text material based on a very large sample.

A	B	C	D	E	F	G	H	I	J
8.2	1.4	2.8	3.8	13.0	3.0	2.0	5.3	6.5	0.1

K	L	M	N	O	P	Q	R	S	T
0.4	3.4	2.5	7.0	8.0	2.0	0.1	6.8	6.0	10.5

U	V	W	X	Y	Z
2.5	0.9	1.3	0.2	2.0	0.07

Similar activities can be constructed for typical word and sentence length. Valuable experience and insight can be gained by applying these results to new settings.

A computer printer prints out 80 characters to the line and 55 lines to the page. How many E's would you expect in 6 full pages, ignoring spaces and punctuation?

$$(6 \times 80 \times 55 \times .13 = 3432)$$

A long sentence contains 26 E's. Estimate the number of letters in the sentence.

$$(26/.13 = 200)$$

A short essay contains 1329 words. How many E's would you expect?

(insufficient information)

Have students sample some material for an average word length and let them apply it to answer the last question above.

Surveying Preferences

Let each student ask 10 different people for their color preference from a selected list, record the results, and display the data in a circle graph using percents. Then have the students pool the results for the entire class and have each draw another circle graph for the total of all those sampled.

How many people in a group of 72 would you expect to prefer the color red?

Simulating Baseball

Batting averages of baseball players are readily available statistics and can serve as the basis for an interesting simulation activity. Select a player from one of your class' favorite teams and simulate what his performance might be for the next few times at-bat.

Here is an example where the simulation is done three different ways on Babe Ruth's record for 2503 games over 22 years.
Compute the probabilities from the batting record.

	AB	H	2B	3B	HR
RECORD	8399	2873	506	136	714
PROBABILITIES		.342	.060	.016	.085

Using dice—Find the closest fractional approximation in terms of 36ths and assign dice outcomes that match these values. One such assignment is shown here but others are possible. Repeated rolls of the dice simulate successive times at-bat.

BATTING SIMULATION	DICE OUTCOMES	PROBABILITY
single	9 or 10	$P(9 \text{ or } 10) = 7/36$
double	11	$P(11) = 2/36$
triple	12	$P(12) = 1/36$
home run	4	$P(4) = 3/36$
out	all other results	23/36

Using this method, probabilities cannot be matched exactly nor can outcomes always be conveniently assigned. But it is an easy way to simulate some additional at-bats based upon the records supplied.

Using a random number table—Assign intervals over 000 to 999 to match the probabilities. Then repeatedly select 3-digit numbers from a random number table to simulate successive at-bats. A sample random number table along with a program for generating and printing one with a microcomputer can be found on page 269.

BATTING SIMULATION	RANDOM NUMBER OUTCOME	PROBABILITY
single	000–180	.181
double	181–240	.060
triple	241–256	.016
home run	257–341	.085
out	342–999	.658

Using a microcomputer—Assign similar intervals as above but using decimals between 0 and 1 to match the probabilities of the various outcomes. Then repeatedly loop through a program that selects random numbers between 0 and 1 to simulate the successive at-bats. Key statements are given here.

```
LET  X = RND(1)          selects a random number less than 1
IF X < .181 THEN ---     single
IF X < .241 THEN ---     double
IF X < .257 THEN ---     triple
IF X < .342 THEN ---     home run
```

In these simulations the probabilities remain fixed. This is not unreasonable since a great many at-bats occurred and only a few additional ones will be simulated. Of course, a program can easily modify the probabilities to reflect each additional at-bat.

8.5 GAMES OF CHANCE

It was the study of games of chance that first led Pascal and Fermat to the invention of probability in the mid-1600s. So it is not surprising that probability plays a role in familiar card and dice games as well as other games of chance.

Poker

Interesting classroom discussion can come from counting the different possible poker hands and computing various related probabilities, such as the following.

How many different 5-card poker hands can be dealt from a deck of 52 cards? The answer is the combination of 52 things taken 5 at a time:

$$_{52}C_5 = \binom{52}{5} = \frac{52!}{5!47!} = 2,598,960$$

This may not seem like very many. However, if you were able to deal out a different 5-card poker hand every second, working day and night, it would take about one month to deal them all.

Why is a flush better than a straight and a straight better than three of a kind in poker? The ranking of the poker hands is based upon the probabilities of their occurring—the better the hand, the lower the probability of its being dealt. These are the probabilities of the various hands:

straight flush	0.0000154	(1 in 64,974 hands)
four of a kind	0.0002401	(1 in 4,165 hands)
full house	0.0014406	(1 in 694 hands)
flush	0.0019654	(1 in 509 hands)
straight	0.0039246	(1 in 256 hands)
three of a kind	0.0211285	(1 in 48 hands)
two pair	0.0475390	(1 in 21 hands)
one pair	0.4225690	(1 in $2\frac{1}{2}$ hands)
no pair	0.5011774	(1 in 2 hands)

Here is a detailed solution for the probability of a fairly common poker hand, one pair (the probabilities of the other hands can be found in much the same way):

$$\text{Probability of one pair} = \frac{\binom{13}{1}\binom{4}{2}\binom{12}{3}4^3}{\binom{52}{5}} = \frac{13 \cdot 6 \cdot 220 \cdot 64}{2{,}598{,}960} = 0.4225690$$

Roulette

Although it is quite probable that very few students have played roulette, it is a game that most have heard of. Bring a small roulette wheel into class and use it to introduce the concept of *mathematical expectation.*

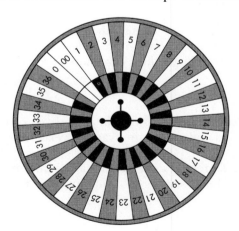

In games of chance, mathematical expectation can be thought of as the weighted mean of possible winnings, each weighted by its probability. Consider a $1 bet on a specific number on the roulette wheel.

The probability of winning is 1 out of 38. If you win, you get $35 plus your $1 back.

The probability of losing is 37 out of 38, and the loss is your $1.

$$E[X] = \frac{1}{38}(35) + \frac{37}{38}(-1) = \frac{35}{38} - \frac{37}{38} = -\frac{1}{19}$$

Your expected winnings in dollars are $-\frac{1}{19}$, or about a 5-cent loss on a $1 bet. Obviously, the game is weighted slightly in favor of the house.

If you're a conservative gambler, you might choose to bet on a color (red or black) instead of on a number. With 18 red and 18 black positions, surely your chances of winning are higher. If you win, you get only $1 plus your $1 back. However, one often forgets the two additional positions, 0 and 00, which are neither red nor black. Surprisingly, the expectation is exactly the same, a loss of about 5 cents per $1 bet.

$$E[X] = \frac{18}{38}(1) + \frac{20}{38}(-1) = \frac{18}{38} - \frac{20}{38} = -\frac{1}{19}$$

Craps

A detailed discussion of the mathematics behind the dice game called craps can shed substantial light on the nature of conditional probability. The game is played with a pair of dice under the following rules.

If a 7 or 11 is rolled, you win.
If a 2, 3, or 12 is rolled, you lose.
If a 4, 5, 6, 8, 9, or 10 is rolled, you roll the dice again until a 7 occurs and you lose, or until the initial score occurs again and you win.

It is this third case that creates the conditional probability situation. Suppose the first roll is a 4 which has 3 chances out of 36 of occurring. Then the only subsequent roll that will end the game is a 4 or a 7. Of the 9 ways to roll either a 4 or a 7, only 3 are 4's. So having shown a 4 on the first roll, the chances of winning the game with another 4 are 3 out of 9. All other scores that may occur at this point are simply ignored.

Win on a 7 or 11	$P(7 \text{ or } 11) = 6/36 + 2/36$	$= 8/36$
Win on a 4	$P(4) \times P(4 \text{ given } 4 \text{ or } 7) = 3/36 \times 3/9$	$= 9/324$
Win on a 5	$P(5) \times P(5 \text{ given } 5 \text{ or } 7) = 4/36 \times 4/10$	$= 16/360$
Win on a 6	$P(6) \times P(6 \text{ given } 6 \text{ or } 7) = 5/36 \times 5/11$	$= 25/396$
Win on an 8	$P(8) \times P(8 \text{ given } 8 \text{ or } 7) = 5/36 \times 5/11$	$= 25/396$
Win on a 9	$P(9) \times P(9 \text{ given } 9 \text{ or } 7) = 4/36 \times 4/10$	$= 16/360$
Win on a 10	$P(10) \times P(10 \text{ given } 10 \text{ or } 7) = 3/36 \times 3/9$	$= 9/324$

Add these separate probabilities to find the probability of winning on the first or any subsequent roll. The chances are just under 1/2.

$$P(\text{win}) = 8/36 + 9/324 + 16/360 + 25/396 + 25/396$$
$$+ 16/360 + 9/324$$
$$= 244/495$$
$$= 0.4929$$

Lotteries

With their increasing popularity, state lotteries offer the basis for some interesting counting and probability problems. In a PICK 6 lottery, 6 numbers from 1 through 42 are chosen at random. The probabilities of guessing 6, 5, or 4 correct are given here. Note how combinations are applied in each case.

$$P(6 \text{ correct}) = \binom{6}{6}\binom{36}{0} \bigg/ \binom{42}{6} = 1/5{,}245{,}786 \quad = 0.0000001$$

$$P(5 \text{ correct}) = \binom{6}{5}\binom{36}{1} \bigg/ \binom{42}{6} = 216/5{,}245{,}786 \quad = 0.0000411$$

$$P(4 \text{ correct}) = \binom{6}{4}\binom{36}{2} \bigg/ \binom{42}{6} = 9450/5{,}245{,}786 = 0.0018014$$

8.6 MORE AIDS AND ACTIVITIES

Most units in mathematics can be enhanced and enriched with probability applications. Some examples are given in this section. Notice how they require insight into and understanding of both the subject at hand and probability. Many also require good problem-solving analysis.

Factors and Primes

1. A factor of 60 is chosen at random. What is the probability
   that it is 20? (1/12)
   that is has factors of both 2 and 5? (1/3)
2. A counting number from 1 through N is chosen at random. What is the probability that it is prime,
   if $N = 10$? (2/5)
   if $N = 50$? (3/10)

Fractions and Decimals

The numbers 3, 4, and 5 are placed on three cards and then two cards are chosen at random.

3. The two cards are placed side-by-side with a decimal point in front. What is the probability that the decimal is more than 3/8? (2/3)
4. One card is placed over the other to form a fraction. What is the probability that the fraction is less than 1.5? (5/6)
5. Repeat the same two questions but starting with four cards numbered 3, 4, 5, and 6. (3/4, 3/4)

Polygons

6. A vertex of a paper isosceles triangle is chosen at random and folded to the midpoint of the opposite side. What is the probability that a trapezoid is formed? (1/3)

7. A vertex of a paper square is folded onto another vertex chosen at random. What is the probability that a triangle is formed? (1/3)

8. Three randomly chosen vertices of a regular hexagon cut from paper are folded to the center of the hexagon. What is the probability that an equilateral triangle is formed? (1/10)

Measurement

9. A piece of string is cut at random into two pieces. What is the probability that the shorter piece is less than half the length of the longer piece? (2/3)

10. A paper square is cut at random into two rectangles. What is the probability that the larger perimeter is more than 1½ times the smaller? (2/5)

Equations

The numbers 2, 3, and 4 are substituted at random for a, b, and c in the equation $ax + b = c$.

11. What is the probability that the solution is negative? (1/2)

12. If c is not 4, what is the probability that the solution is negative? (3/4)

Algebra

The numbers 1, 2, and 3 are substituted at random for a, b, and c in the quadratic equation $ax^2 + bx + c = 0$.

13. What is the probability that $ax^2 + bx + c$ can be factored? (1/3)

14. What is the probability that $ax^2 + bx + c = 0$ has real roots? (1/3)

Geometry

15. Two faces of a cube are chosen at random. What is the probability that they are in parallel planes? (1/5)

16. Three edges of a cube are chosen at random. What is the probability that each edge is perpendicular to the other two? (2/55)

17. A point P is chosen at random in the interior of square ABCD. What is the probability that triangle ABP is acute?

$$[P(\text{acute triangle}) = 1 - \pi/8 = .6073]$$

Trigonometry

Find the probability for each of these results:

18. The sine of a randomly chosen acute angle is greater than 0.5. (2/3)

19. The cosine of a randomly chosen obtuse angle is greater than -0.5. (1/3)

20. The tangent of a randomly chosen acute angle is greater than 1. (1/2)

8.7 OVERHEAD PROJECTOR IDEAS

The overhead projector can be a powerful teaching tool when used effectively. It can generate new interest and motivate increased attention through its versatility. This section lists various roles the overhead projector can play in the classroom. The examples are from probability and statistics, but the techniques are equally applicable to all areas of mathematics.

To Dramatize Action

Rule a transparency with lines 1 inch apart. Drop pennies on the ruled surface and investigate the probability that a penny will not land on a line. The projected image vividly dramatizes the action for the whole class to see.

Experiment first and compute an empirical probability. Then analyze the problem geometrically for a theoretical probability.

The diameter of a penny is 3/4 inch. If its center point falls in the quarter-inch strips in the middle of the one-inch spaces between the lines, the penny itself will not cross a line. This is 1/4 of the total area and so the probability of not falling on a line is 1/4.

To Focus Attention

Place the first few rows of Pascal's triangle on a transparency. Then cut some strips of colored acetate to focus attention on specific rows and columns and their special properties.

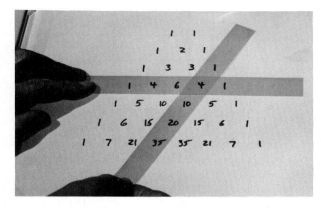

The strip over row 4 shows all the possible combinations of 4 things.

$$\binom{4}{0} = 1 \qquad \binom{4}{1} = 4 \qquad \binom{4}{2} = 6 \qquad \binom{4}{3} = 4 \qquad \binom{4}{4} = 1$$

The other strip shows the first few possible combinations taken 3 at a time.

$$\binom{3}{3} = 1 \qquad \binom{4}{3} = 4 \qquad \binom{5}{3} = 10 \qquad \binom{6}{3} = 20 \qquad \binom{7}{3} = 35 \ldots$$

The two strips cross at the combination of 4 things taken 3 at a time.
Notice that the sum of the numbers in row 4 is 16.

$$1 + 4 + 6 + 4 + 1 = 16$$

Move the strip to some other rows and see if the students can discover and then generalize this property:

The sum of the numbers in row n of Pascal's triangle in 2^n.

To Supply Grids

Project a 4 x 4 grid on the board and let it represent a 16-block area. Start at point P and take a random walk of 4 blocks. Determine your path by tossing a coin 4 times.

If the coin falls heads (H), walk one block to the right.
If the coin falls tails (T), walk one block up.

As a coin is repeatedly tossed, trace the path on the projected grid for the class. What is the probability that a four-block random walk takes you to the centerpoint C?

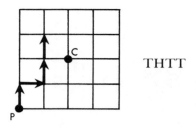

THTT

The path shown comes from the four tosses THTT. Experiment a few times tracing out on the projected grid the four-block walks that result. Then analyze the problem mathematically.

There are $2^4 = 16$ ways the coins can fall but only $\binom{4}{2} = 6$ of them will take you to point C. Hence the required probability is 6/16 or 3/8. Note the connection between this problem and the five numbers in row 4 of Pascal's triangle.

To Present Facts

Facts that are detailed or extensive can be prepared in advance on an acetate and projected at the appropriate time during the class for the students to use. These facts from a recent year illustrate such data. They will be used in examples that follow.

TRAFFIC FATALITIES PER 100 MILLION MILES BY STATES

AL	6.4	HI	4.7	MA	3.5	NM	8.0	SD	5.4
AK	8.8	IO	7.1	MI	4.2	NY	4.7	TN	7.1
AZ	6.2	IL	4.3	MN	4.6	NC	6.2	TX	5.2
AR	5.6	IN	5.1	MS	5.6	ND	4.8	UT	5.5
CA	4.4	IA	5.9	MO	5.6	OH	4.5	VT	4.7
CO	5.3	KS	5.0	MT	7.0	OK	5.0	VA	4.5
CT	2.8	KY	5.6	NE	4.4	OR	5.3	WA	4.3
DE	5.2	LA	7.1	NV	8.0	PA	4.1	WV	6.2
FL	5.5	ME	4.6	NH	4.6	RI	3.0	WI	4.7
GA	6.1	MD	3.9	NJ	3.2	SC	6.5	WY	6.5

To Display Results

Here is one way to illustrate how to draw a histogram on the overhead projector for the traffic fatality data given above. Start with a centimeter grid copied on a transparency and 50-cm cubes. After labeling the axes, put a cube in the appropriate place for each state. This demonstration is convenient and easy to perform even with student participation.

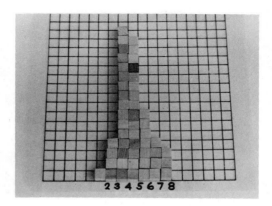

Using a tabulation method called a *stem-and-leaf plot,* a similar representation that is both numerical and visual can be made quickly and easily.

```
2 | .8
3 | .0 .2 .5 .9
4 | .1 .2 .3 .3 .4 .4 .5 .5 .6 .6 .6 .7 .7 .7 .7 .8
5 | .0 .0 .1 .2 .2 .3 .3 .4 .5 .5 .6 .6 .6 .6 .9
6 | .1 .2 .2 .2 .4 .5 .5
7 | .0 .1 .1 .1
8 | .0 .0 .8
```

The units digit forms the stem and the tenths digit the leaf for each entry in this plot. The stem 4 has sixteen leaves while the stem 2 has only one.

Once the stem-and-leaf plot is complete on acetate, rotate it 90° counterclockwise to show how the tabulation also forms a graph similar to the histogram. However, it has the distinct advantage that none of the original data are lost in presenting the graph.

To Show Motion or Change

When using the normal distribution, probabilities are found as areas under the curve. Their specific values can be found from tables but they can be vividly represented using this idea.

Cut a carefully drawn normal distribution from a piece of heavy paper or oaktag. A side of a file folder will do nicely. Then place a sheet of colored acetate over it in the appropriate position. The colored acetate can be moved about to show various areas. For some problems, two sheets of colored acetate will be needed.

The shaded region projected here represents the probability that a score lies more than 1 standard deviation above the mean.

$$P(Z > 1) = 0.1587$$

8.8 CALCULATOR AND COMPUTER APPLICATIONS

The calculator allows for the handling of real data from published reference sources and from classroom experimentation without the tedious aspects of hand computation. Hence, the focus of such work can be problem-solving as currently recommended. The microcomputer allows for convenient and interesting simulations that are both fast and accurate and make an excellent follow-up to the hands-on aspects of statistical experimentation in the classroom.

Numerous examples have already been given where the calculator would be a useful tool. Probability and statistics also offer an opportunity to discuss and review procedures possible on a calculator and hence can be useful in developing greater computational skill with these devices.

As an example consider these data on heights and arm spans collected from 29 students. The comments on page 257 indicate the close relationship expected between these two measurements. Students can profit from a detailed discussion of the computational procedures required on a calculator in computing the sample means and standard deviations. Both the formula for the definition of the standard deviation given first and the equivalent computational form that follows yield the same results. Discuss which one is more convenient to use with a calculator.

HEIGHTS AND ARM SPANS
IN INCHES

HEIGHT	f	ARM SPAN	f
56	3	56	4
58	6	58	6
60	7	60	7
62	4	62	2
64	1	64	2
66	5	66	2
68	3	68	4
70	0	70	2
	29		29

$$\overline{x} = \frac{\Sigma x}{n}$$

$$s = \sqrt{\frac{\Sigma(x - \overline{x})^2}{n - 1}} = \sqrt{\frac{\Sigma x^2 - \dfrac{(\Sigma x)^2}{n}}{n - 1}}$$

	MEAN	ST. DEV.
HEIGHTS	61.45	3.85
ARM SPANS	61.66	4.54

Randomness is a very important concept in any discussion of probability. While the computer can repeatedly simulate the toss of a coin or the roll of a die, students should begin with the actual, concrete hands-on experiences. These should be followed by a discussion of random number tables where students can visually scan, tabulate, and compare in exploring the nature of randomness. Have the students use the tables to simulate a variety of activities.

Here is a BASIC program that can be used to generate and print out random number tables for classroom use. Repeated runs will produce different tables.

```
10   PRINT "1000 COMPUTER-GENERATED RANDOM DIGITS": PRINT : PRINT
20   FOR I = 1 TO 20
30   FOR J = 1 TO 50
40   PRINT  INT (10 * RND (1));
50   IF J / 5 = INT (J / 5) THEN PRINT " ";
60   NEXT J
70   IF I / 5 = INT (I / 5) THEN PRINT
80   PRINT : NEXT I
999  END
```

```
]RUN
1000 COMPUTER-GENERATED RANDOM DIGITS

51931 49880 20265 39125 76493 53962 33771 59226 83098 27326
21778 93845 25889 40816 61165 48729 23683 31693 77752 56238
00348 77577 82005 82615 40791 31616 80349 76090 29995 68782
31677 17503 61901 25545 36207 04852 21476 17837 24206 58804
75349 94418 19699 97660 34193 83480 95051 64990 90907 15420

39270 89692 63491 13042 09509 32802 04895 06389 81574 96730
08721 53044 60762 00920 56319 31648 48081 05479 97157 16480
99256 81023 83761 42819 48098 18051 18952 15369 10593 28431
41308 56422 61946 63556 43493 58162 69038 43880 98199 39926
35830 19005 39525 63934 06870 02054 76766 93195 76674 50628

33668 07271 48510 62488 68375 70036 53899 50149 46532 29511
02984 59710 92837 08179 72719 47423 85542 53916 52507 21460
45692 59934 06988 94706 94923 54350 25766 33446 35135 12816
09876 61503 08961 57617 71349 35178 80400 73104 29055 25400
01396 50574 45755 08687 93916 27152 45859 44751 83960 43526

45816 01060 42257 47798 49132 24623 53505 86103 93601 54600
27018 06459 01399 36183 14008 87707 21230 93781 15359 97561
41176 76142 33335 85780 78759 39359 11359 81872 51239 11015
08029 54342 66420 50385 26989 68252 87891 04981 13446 56363
75914 17846 59997 95088 06807 83775 82904 32471 39756 22999
```

This program simulates a set of 10 rolls of a pair of dice. The number of rolls can easily be modified as desired. Producing this program can be a useful programming exercise for students following a hands-on classroom activity.

```
10   REM  ROLLING DICE
20   FOR N = 1 TO 10
30   LET A =  INT (6 * RND (1) + 1)
40   LET B =  INT (6 * RND (1) + 1)
50   LET S = A + B
60   PRINT S;" ";
70   NEXT N
999  END

]RUN
2 9 10 9 6 11 10 5 9 3
]RUN
7 9 9 2 7 11 11 7 6 11
]RUN
3 11 7 7 3 5 9 5 3 7
]RUN
8 8 7 6 5 5 9 8 7 7
```

```
]RUN
7 6 9 9 11 10 6 8 7 5
]RUN
9 7 8 7 9 7 8 8 7 8
```

The 6 runs show the simulated results of 60 random rolls of a pair of dice. They can be very useful in discussing the nature of randomness. No 4's occurred and yet 5 were expected. Fifteen 7's occurred and yet only 10 were expected. Many would say that the last run is certainly not a random result but it is. On the other hand, it is a very unexpected result since only 7's, 8's, and 9's occurred.

An even more valuable experience is to compare experimental results to much larger computer simulated results. Here is the output of a modified version of this program designed to count the number of tosses that result for each sum when 3600 was entered as input for N, the total number of tosses.

SIMULATED		EXPECTED
107	TWOS	100
227	THREES	200
279	FOURS	300
382	FIVES	400
489	SIXES	500
562	SEVENS	600
488	EIGHTS	500
412	NINES	400
315	TENS	300
234	ELEVENS	200
105	TWELVES	100
3600	TOTAL	3600

A computer simulation can also be useful as a follow-up to a theoretical analysis of a problem. Consider the polyhedron dice problem on page 248. The theoretical probability of rolling a 7 with the five regular polyhedrons numbered 1 to 4, 6, 8, 12, and 20 respectively was computed to be 1 in 3072. Input 3072 for N in this program, run it a few times, and see how many 7's result on each try. Pool the results and discuss again what is meant by probability in the long run.

```
10      REM     POLYHEDRON DICE
20      LET X = 0
30      INPUT "NUMBER OF ROLLS "; N
40      FOR Y = 1 TO N
50      LET T = INT(4 * RND(1) + 1)
60      LET H = INT(6 * RND(1) + 1)
70      LET O = INT(8 * RND(1) + 1)
80      LET D = INT(12 * RND(1) + 1)
90      LET I = INT(20 * RND(1) + 1)
```

```
100     LET S = T + H + O + D + I
110     IF S < > 7 THEN 130
120     LET X = X + 1
130     NEXT Y
140     PRINT
150     PRINT   N; " ROLLS"
160     PRINT   X; " SEVENS"
170     PRINT   "PROBABILITY OF A SEVEN "; X/N
999     END
```

Computer simulations can be used to find areas under a curve. When applied to a unit circle, this simulation can be used to approximate π.

```
10      REM     APPROXIMATION FOR PI
100     LET H = 0
110     INPUT "NUMBER OF TOSSES "; T
120     FOR I = 1 TO T
130     LET X = RND(1)
140     LET Y = RND(1)
150     IF Y > SQR(1 - (X^2)) THEN 170
160     LET H = H + 1
170     NEXT I
200     LET A = 4 * H/T
210     PRINT "APPROXIMATION FOR PI: "; A
999     END
```

Imagine darts randomly thrown in a unit square in the first quadrant with a vertex at the origin. This program simulates that process and counts those hitting inside the quarter-circle with radius 1 and center at the origin. The ratio of the number of hits (H) to the total number thrown (T) multiplied by 4 gives an estimate of the area of the entire unit circle. Of course, that same value is an estimate of pi since the area of a unit circle is pi.

By adding new lines to the program above, the program can be modified to show the process using the high resolution graphics of the Apple II.

```
15      REM     DRAW CIRCLE
20      HGR
25      HCOLOR = 3
30      HPLOT 140,0 TO 140,159
35      HPLOT 0,80 TO 279,80
40      FOR X = 140 TO 199
45      LET Y = 80 - SQR(3600 - (X - 140)^2)
50      HPLOT X,Y
55      NEXT X
60      HPLOT 140,20 TO 200,20 TO 200,80
95      REM     THROW DARTS
145      HPLOT 60 * X + 140, 80 - 60 * Y
180      REM     COMPUTE APPROXIMATION
190      PRINT T;" DARTS;    ";H;" HITS"
```

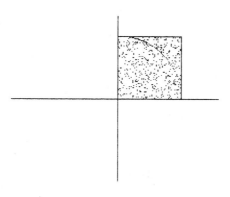

```
]RUN
NUMBER OF TOSSES 500
500 DARTS; 395 HITS
APPROXIMATION FOR PI: 3.16
```

The microcomputer can be a valuable problem-solving tool for situations that require a great deal of repeated computation with changing values of the variables. Consider the birthday problem discussed in this chapter on page 254. With 365 choices possible, the probability of at least one duplication is about 1/2 with a group of 23 people.

In general, this formula can be used to find the probability of at least one duplication among N choices with M people. Note that there are M factors in both the numerator and the denominator of the fraction.

$$P(\text{at least one duplication}) = 1 - \frac{N(N-1)(N-2)\dots(N-M+1)}{N(\ N\ )(\ N\ )\dots(\ \ \ N\ \ \ )}$$

For small values of N and M, this probability can be computed with pencil and paper or with a calculator. Here is a program written in BASIC that will tabulate the probabilities for any values of N. With M taking on the value of N + 1, the last entry will always have a probability of 1. That is, a duplication is certain to happen with N choices if there are N + 1 people.

```
10   REM   DUPLICATION PROBABILITIES
20   PRINT
30   INPUT "NUMBER OF CHOICES ";N
40   PRINT
50   PRINT "NUMBER OF","PROBABILITY OF"
60   PRINT "PEOPLE","AT LEAST ONE DUPLICATION "
70   PRINT
80   LET P = 1
90   FOR A = 1 TO N + 1
100  LET P = P * (N - A + 1) / N
110  LET D = 1 - P
120  PRINT A,D
130  NEXT A
999  END
```

```
JRUN

NUMBER OF CHOICES 10

NUMBER OF          PROBABILITY OF
PEOPLE             AT LEAST ONE DUPLICATION

    1              0
    2              .1
    3              .28
    4              .496
    5              .6976
    6              .8488
    7              .93952
    8              .981856
    9              .9963712
   10              .99963712
   11              1
```

The program can be changed to list probabilities for various numbers of choices for a given number of people. If you have a party with 10 people, here are some of the probabilities for at least one duplication. Write a program that will produce these results when you input a value of 10 for the number of people (M).

```
NUMBER OF PEOPLE 10

NUMBER OF          PROBABILITY OF
CHOICES            DUPLICATION

    10                .99963712
    20                .934527092
    30                .815361225
    40                .706650684
    50                .61829332
    60                .547532414
    70                .490376101
    80                .443554267
    90                .404635448
   100                .371843491
```

This table gives the least number of people needed for a probability greater than $\frac{1}{2}$ for some special values of N.

	NUMBER OF CHOICES	LEAST NUMBER OF PEOPLE FOR A PROBABILITY GREATER THAN $\frac{1}{2}$
For months of the year:	12	5 (.618055556)
For letters of the alphabet:	26	7 (.587227296)
For cards in a deck:	52	9 (.519745295)
For days of the year:	365	23 (.507297234)

EXERCISES

1. Suppose a person is chosen at random from the entire world's population. What is a reasonable estimate of the probability that the person would come from the United States?

2. A special set of dominoes has faces that are numbered from 0 through 9. How many are in the complete set?

3. A factor of 72 is chosen at random. What is the probability that it does not have a factor of 3?

4. Three cards are numbered 4, 5, and 6 respectively. If they are randomly arranged to form a 3-digit number, what is the probability that it will be divisible by 4? by 3?

5. Two six-sided dice are numbered 1, 1, 3, 3, 5, 5 and 2, 2, 4, 4, 6, 6 respectively. What is the probability that, when rolled, they will show a sum of 7?

6. What is the probability of rolling a sum of 7 using three regular dice?

7. In how many different ways can the nine letters in the word "SEVENTEEN" be arranged?

8. The numbers 2, 3, and 5 are substituted at random for a, b, and c in the quadratic $ax^2 + bx + c$. What is the probability that the resulting expression can be factored?

9. Find the probability that a randomly chosen acute angle has a tangent less than 1.

10. Two randomly chosen vertices of a paper square are folded to the center of the square. What is the probability that a hexagon is formed?

11. Two edges of a cube are chosen at random. What is the probability that they lie in two lines that are skew?

12. Two regular dice are rolled and the numbers on their faces multiplied. Draw a stem-and-leaf plot of all possible products on the roll of a pair of regular dice.

13. If the letters in a newspaper article contain 315 T's, estimate how many letters were used in all.

14. Pennies are randomly tossed on a grid ruled in 2-inch squares. What is the probability that a penny will fall on a line?

15. Suppose you ask individuals for their random choices of letters of the alphabet.

How many people would you need to ask so that the probability of at least one duplication becomes better than 1 in 2? Use trial and error with your calculator. What assumption is being made?

16. Show how to set up and compute the probability of being dealt a full house—three of one kind and two of another—in a 5-card poker hand dealt from a regular deck of 52 cards.

ACTIVITIES

1. Sample 10 people for their estimates of the diameter of a quarter from these choices in inches.

$$3/4 \qquad 13/16 \qquad 7/8 \qquad 15/16 \qquad 1$$

Draw a histogram of the results. Which is the correct choice?

2. Copy down the 8-digit serial number from a dollar bill. Create and answer five probability questions that relate to possible arrangements of those digits.

3. A photograph is cut into four strips of equal size and the pieces are rearranged at random. Describe how this activity could be used in a classroom to motivate discussion on a probability problem.

4. The eight letters in the word "NINETEEN" are placed on separate cards. How could they be used as a teaching aid in developing a method for finding the probability that a random arrangement of the letters has all three E's together? no two E's together?

5. Repeat the coin-tossing activity described on page 249. Include in the table the cumulative percents and draw a line graph of the results.

6. Experiment with Buffon's needle problem using at least 200 drops.

7. Ask 23 people selected at random and see if you find at least one duplication in birth dates.

8. Draw a 35° angle on a piece of paper and ask 15 people to estimate its measure in degrees. Compute the mean for the results. Do your sample results imply that people tend to overestimate, underestimate, or accurately estimate its size?

9. Sample 500 words from some ordinary newspaper text material. Find the mean and the standard deviation for the word length used.

10. A student showed this work in computing the probability of two pairs in a poker hand. Describe how you would use a deck of cards to identify the error for the student and give the correct method for finding the answer.

$$\binom{13}{1}\binom{4}{2}\binom{12}{1}\binom{4}{2}\binom{44}{1}\Big/\binom{52}{5}$$

11. Find a source of numerical data that could be used effectively in the classroom as an example to demonstrate a stem-and-leaf plot. Supply the completed plot.

12. Simulate 36 rolls of a pair of dice by using the random number table on page 270. Compare your results with those expected.

13. Set up arbitrary probabilities and describe how you can use a random number table to simulate a bowling game.

14. Run the computer simulation described on page 272 to approximate pi. Use T = 500.

15. Design and construct a transparency of your own that can be used in teaching probability.

16. Investigate the uses of probability by insurance companies in establishing their mortality tables.

17. Prepare a lesson introducing probability to a first-year general mathematics class.

READINGS AND REFERENCES

1. Read the 1981 Yearbook of the National Council of Teachers of Mathematics titled *Teaching Statistics and Probability.* Prepare a report on two chapters that you find especially useful.

2. Read *How to Lie with Statistics* by Huff and Geis, New York: W. W. Norton and Company, Inc., 1964. Summarize three misuses of statistics presented therein that are suitable for discussion in a secondary classroom.

3. A wide variety of applications of statistics can be found in *Statistics: A Guide to the Unknown* edited by Judith Tuner, San Francisco: Holden-Day, Inc., 1972. Summarize five of the applications found in this report by the Joint Committee on the Curriculum in Statistics and Probability of the American Statistical Association and the National Council of Teachers of Mathematics.

4. An interesting collection of statistical ideas and their relevance in today's society can be found in *Statistics: Concepts and Controversies* by David Moore, San Francisco: W. H. Freeman and Company, 1979. Identify three specific examples that would be useful in teaching statistics.

5. Read Chapter 5 titled "Greatness and Misery of Man" in *Men of Mathematics* by E. T. Bell, New York: Simon and Schuster, 1965. Prepare a report on the life of the mathematician described therein.

6. Read and summarize Chapters 22 and 23 on statistics and probability in *Mathematics in Western Culture* by Morris Kline, New York: Oxford University Press, 1953.

7. Read Chapter 7 titled "Chance and Chanceability" in *Mathematics and the Imagination* by Edward Kasner and James Newman, New York: Simon & Schuster, 1967. Outline the key topics covered.

8. Read the article titled "Buffon's Needle Problem on a Microcomputer" in the November 1981 issue of the *Mathematics Teacher* of the National Council of Teachers of Mathematics. Run the program presented there.

9. Read and summarize two articles from the "Activities" section of the *Mathematics Teacher* of the National Council of Teachers of Mathematics that deal with probability and statistics. Those listed below also can be found in the NCTM publication titled *Activities for Implementing Curricular Themes from the Agenda for Action* edited by Christian Hirsch, 1986.

> "Data Fitting without Formulas" by Shulte and Swift, April 1986
> "Plotting and Predicting from Pairs" by Shulte and Swift, September 1984
> "Stem-and-Leaf Plots" by Landwehr and Watkins, October 1985

10. The National Council of Teachers of Mathematics publication, *Teaching with the NCTM Student Math Notes*, 1987 has numerous articles devoted to probability and statistics. Read two that deal with these topics, summarize their content, and describe the level at which they would be most appropriate. Several are listed here:

"Correlation: What Makes a Perfect Pair?"
"Duplication Probabilities"
"Random Walks"
"Successful Simulation"

11. Read Chapter 6 titled "Probability and Statistics" in *Using Computers in Mathematics* by Gerald Elgarten, Alfred Posamentier, and Stephen Moresh, Menlo Park, California: Addison-Wesley Publishing Company, 1983.

Answers to Selected Exercises and Activities

CHAPTER 1 (PAGE 23)

Exercises

1. .93 miles   **2.** Approximately 95 miles   **3.** 20   **4.** 25
5. Assume the distance is 3045 miles and the length of a pencil is 7.5 inches. Then approximate 5280 as 5000 and 12 as 10 to obtain this estimate:

$$3000 \times 5000 \times 10 \times \frac{2}{15} = 20,000,000$$

   The exact number required is 25,724,160.
6. One billion seconds is approximately 32 years. Find a historical event that occurred about 32 years ago.
7. $\frac{n}{2}(n + 1)$ **8.** $n = 41$ **11.** 90   **17.** The sum of the three numbers is 12.
18. For the first sequence, consider the products $0 \times 1, 2 \times 3, 4 \times 5, 6 \times 7, \ldots$; the next two numbers are $8 \times 9 = 72$ and $10 \times 11 = 110$. The second sequence is 12, 1, 1, 2, 1, 3, 1, ...; these represent the number of chimes on a clock, starting at noon, that strikes once on the half hour and the number of hours on the hour.
19. Start in the lower left hand corner and go up, right, down, left, etc., in a spiral manner, following the pattern hexagon, circle, square, square, circle, circle.
20. Draw a circle with center B and radius equal to the distance AB. From A, use the radius of the circle and mark off three arcs. The third one will determine the required point C.

Activities

19. This LOGO program draws the first five stages of the snowflake. Use input values of 180 for X and 0 for Y.

```
TO SNOW :X :Y
FULLSCREEN
HT
PU
SETPOS [-80 -40]
PD
RT 30
REPEAT 3[FD :X FLAKE :X :Y RT 120]
IF :Y = 4 [STOP]
CS
PU HOME PD
SNOW :X/3 :Y+1
END


TO FLAKE :X :Y
IF :Y = 0 [STOP]
FLAKE :X :Y-1
LT 60 FD :X
FLAKE :X :Y-1
RT 120 FD :X
FLAKE :X :Y-1
LT 60 FD :X
FLAKE :X :Y-1
END
```

CHAPTER 2 (PAGE 49)

Exercises

2. 2,097,151 **3.** (d) **4.** (b) **5.** (c) **6.** Almost 1600. **7.** $107,712 **8.** 80,000

9. About 1 week.

10. Begin with 20 and remove n to obtain $20 - n$, or $10 + (10 - n)$. Add the digits in the remaining number of matches to obtain $1 + (10 - n)$ or $11 - n$. Then subtract: $(20 - n) - (11 - n) = 9$.

13. Let a, b, and c represent the numbers showing on the dice. The algebraic steps are as follows: $2a$, $2a + 5$, $10a + 25$, $10a + b + 25$, $100a + 10b + 250$, $100a + 10b + c + 250$. By subtracting 250 you obtain the three digit number (representing the three dice) as $100a + 10b + c$.

14. The shortest connecting path for the vertices of a square is shown below. The segments form 120° angles about each of the two interior points. For a square of side s, the total length of the path is $s(1 + \sqrt{3})$. For a one-inch square, the length is $1 + \sqrt{3}$, or approximately 2.732 inches.

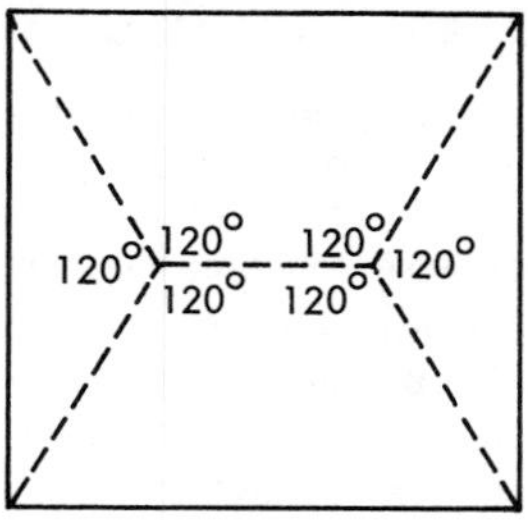

Exercises

1. 19, since there must be 19 losers. **2.** 7 **3.** 0
4. 8; the tens' digits repeat in cycles of 20.
5. 8; try solving a smaller related problem, such as finding the 100th or 1000th digit.
6. 10,000 **7.** (a) 610; (b) $\dfrac{(p + q - 2)(p + q - 1)}{2} + p$ **9.** See Ex. 10.
10. The entries in the second row for numbers 0 through 6, 7, 8, and 9 are as follows:

0 through 6	3 2 1 1 0 0 0
0 through 7	4 2 1 0 1 0 0 0
0 through 8	5 2 1 0 0 1 0 0 0
0 through 9	6 2 1 0 0 0 1 0 0 0

11. Call the mathematicians A and B respectively. Then after each of the four statements the following conclusions can be reached: (1) A does not have 16. (2) B does not have 16 or 1. (3) A does not have 8 or 1. (4) B does not have 2 or 8. At this point A knows that B has 4.
12. "You won't give me the penny." **13.** 10
14. 1700 feet; note that when the boats meet for the second time the total distance they will have travelled is three times the width of the river, and $2100 - 400 = 1700$.
16. $74\dfrac{2}{27}$ cu. in.
17. 15
18. Odd, there are an odd number of odd numbers through 1989. Thus after all pairings are completed, there will be a single odd number left.

Activities

9. Isolate the maximum volume and then change line 50 accordingly using first STEP .1 and then STEP .01; 74.073852

Exercises

1. Fill the 13-gallon can and use this to fill up the 5-gallon can, leaving 8 gallons. Pour the 8 gallons into the 11-gallon can. Repeat this procedure but this time leave the 8 gallons in the 13-gallon can. There will also be 8 gallons left now in the 24-gallon can.
3. Weigh any three coins against another set of three. If they balance, go to the remaining three and weigh one against one. If these balance, then the remaining one is counterfeit; if they do not balance, the lighter one is counterfeit. If the original weighing does not balance, select the three on the lighter side and proceed to weigh one against one as above.
4. Begin by weighing four against four. If they balance, weigh two against two. If these balance, weigh one against one to find the lighter coin. In each of the preceding steps, if they do not balance, use the coins on the lighter side of the scale and distribute them in the next weighing.
5. Hint: The key to the solution is the interchanging of coins from one pan to another at one point.
6. $97.62 + 10.62 = $108.24

7. Begin both hour glasses at the same time. When the 7-minute timer is finished, place the egg in the pot. Then there will be four minutes left in the 11-minute timer. Allow this to finish and then turn it over again for another 11 minutes to give the total of 15 minutes.

9. Hint: There are 12 possible figures.

10. Each domino must cover one red and one black square. The two in opposite corners are of the same color. Thus it is not possible to cover the board and leave two like squares uncovered.

11. Draw a coin from the box labeled DP. Assume you draw a D. Then the other coin in the box must also be a D. Therefore this box is DD. Then the box labeled PP cannot contain two pennies and thus must be DP. The third box must be PP. (A similar line of reasoning exists if you assume the first draw is P.)

12. To solve this problem, note that everyone says they are truthtellers.

Activities

7. 3.6.3.6 and 3.3.3.3.6

CHAPTER 5 (PAGE 151)

Exercises

1. 64   **2.** $98 - 67$; 96×87 **3.** $7/4 + 6/5$; $6/5 - 4/7$ **11.** 8

12. There is no last digit.   **13.** 5149   **15.** 1, 6, 15, 28, 45, 66; $n(2n - 1)$

18. $n^2(n^2 + 1)/2$

Activities

3. For n folds, there are $(n + 1)(n + 2)/2$ rectangles.

4. For a 4×4 grid, there are $1^3 + 2^3 + 3^3 + 4^3 = 100$ rectangles of all sizes formed.

CHAPTER 6 (PAGE 190)

Exercises

2. $y = -x^2$ **3.** $w = -l + 10$; $p = 20$; $A = 10l - l^2$ (all for $0 < l < 10$)

5. 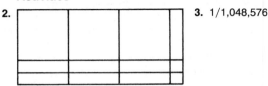   ($1{,}000{,}000 = 4 \times 60^3 + 37 \times 60^2 + 46 \times 60^1 + 40 \times 60^0$)

7. 301; yes

Activities

2. 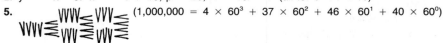   **3.** $1/1{,}048{,}576$

CHAPTER 7 (PAGE 239)

Exercises

1. 10 2. 3/4 sq. in., $2\sqrt{2} + 1$ in.; QRNP, ODNP 3. 84.19 cu. in., 58.57 cu. in. 4. 10

8.

V	4	8	6	20	12
F	4	6	8	12	20
E	6	12	12	30	30

9. 29 11. $x^2/100 + y^2/64 = 1$ 12. The triangles must be acute.

13. $V = 12$, $F = 14$, $E = 24$ 14. 53 1/3 cu. in., $48 + 16\sqrt{3}$ sq. in.

Activities

2. As the cut moves closer and closer to the edge opposite the fold, the ratio approaches 1/2, but it is never equal to 1/2.

4.

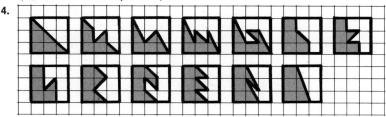

25. Locate points A and B in the shaded region shown in the first figure. Then cut the triangles as indicated. Since these points lie outside the four arcs intercepting diameters, they must form acute angles with the endpoints of the diameters.

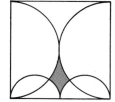
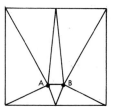

CHAPTER 8 (PAGE 275)

Exercises

1. approximately 1 in 18 2. 45 3. 1/3 4. 1/3; 1 5. 1/3 6. 5/72 7. 7560 8. 1/3
9. 1/2 10. 1/3 11. 4/11 12.

```
0 | 12233444556666889
1 | 00222255688
2 | 00445
3 | 006
```

13. 3000 14. 39/64

15. 7, assuming all letter choices are equally likely 16. $\dfrac{\binom{13}{1}\binom{4}{3}\binom{12}{1}\binom{4}{2}}{\binom{52}{5}} = 0.0014406$

Activities

10. Each pair is counted twice. Use $\dfrac{\binom{13}{2}\binom{4}{2}\binom{4}{2}\binom{44}{1}}{\binom{52}{5}}$

ANSWERS TO SELECTED EXERCISES AND ACTIVITIES   283

Index

O

Octohedron, 205, 209, 221, 248
  truncated, 221
Optical illusions, 41
Overhead projector, 137, 180, 231,
    264
  for axes translation, 180
  for classifying triangles, 233
  for discovering patterns, 182
  to dramatize action, 264
  flow charts for, 137
  to focus attention, 264
  for geoboards, 199
  for graphing absolute value, 180
  for reading a protractor, 232
  for reviewing exponents, 139
  for rotating disks, 139
  to show motion and change, 167
  for sketching prisms, 232
  to supply grids, 265

P

Paper folding, 11, 116, 166
  for algebraic identities, 171, 172
  for arithmetic concepts, 104
  for geometric models, 208
  for geometric properties, 219, 226,
    227
Parabola, 162, 165, 214, 216, 218
  on graph paper, 162
Pascal, Blaise, 45, 159
Pascal's triangle, 250, 264
Pattern discovery, 60, 156, 180
Patterns for a cube, 224
Peg game, 86
Pentagonal:
  numbers, 121, 148
  prism, 209
Percent, 109, 197
Percent chart, 111
Perfect numbers, 146
Perimeter, 196, 197, 198, 200
Permutations, 247
Pi (π), 6, 11, 109, 142, 186, 254, 272
Pick's formula, 199, 202
Plaiting, 209
Plato, 8, 205

Platonic solids, 205
Poker, 259
Polya, George, 28
Polygons:
  angles of, 234
  areas of, 199
  classifying, 196, 212
  recognizing, 196
Polyhedron dice, 248, 271
Polyhedrons:
  construction of, 206–14
  nets for, 209
  regular, 205
  semiregular, 206
  sketching, 203
Polyominoes, 88
Pool, 85
Posters, 43
Powers, 143
Prime numbers, 141, 145, 146, 262
Prism(s):
  construction of, 207
  hexagonal, 203
  pentagonal, 205
  rectangular, 204
  sketching, 203, 232
  triangular, 204
Probability:
  aids for, 243
  in algebra, 263
  in arithmetic, 262
  experimental, 249
  in geometry, 263
  theoretical, 249
  in trigonometry, 263
Probability experiments:
  with birthdays, 254
  with cards, 246
  with coins, 249
  with dice, 247
  with pennies, 253
  with random walks, 251
  with thumb tacks, 252
Problem solving, 53
  strategies for, 54, 57, 60, 63, 65,
    68
Problem of the week, 22
Programs (*See* Computer programs)
Proof, 156
Protractor, 232

Tricks, mathematical, 20, 31
Trigonometry, 263
Trinomials, factoring, 174
Triomino, 88
Truncated:
  cube, 220
  octahedron, 221
  tetrahedron, 211
Truncation, 220

U

Unsolved problems, 8

V

Variables, 164
Visualization activities, 196, 219–23,
  224, 225
Volume, 194, 195, 235

W

Witch of Agnesi, 9
Wolfskehl, Paul, 9
Worksheets (*See* Student worksheets)